D1372145

Sustainability

BUILDING ECO-FRIENDLY COMMUNITIES

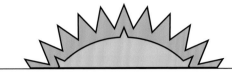

GREEN TECHNOLOGY

Sustainability

BUILDING ECO-FRIENDLY COMMUNITIES

Anne Maczulak, Ph.D.

Facts On File
An imprint of Infobase Publishing

SUSTAINABILITY: Building Eco-Friendly Communities

Copyright © 2010 by Anne Maczulak, Ph.D.

All rights reserved. No part of this book may be reproduced or utilized in any form or by any means, electronic or mechanical, including photocopying, recording, or by any information storage or retrieval systems, without permission in writing from the publisher. For information contact:

Facts On File, Inc.
An imprint of Infobase Publishing
132 West 31st Street
New York NY 10001

Library of Congress Cataloging-in-Publication Data

Maczulak, Anne E. (Anne Elizabeth), 1954–
    Sustainability: building eco-friendly communities / Anne Maczulak.
        p. cm.—(Green technology)
    Includes bibliographical references and index.
    ISBN-13: 978-0-8160-7201-9
    ISBN 10: 0-8160-7201-9
    1. Sustainable development. 2. Sustainable living. 3. Green technology. I. Title.
    HC79.E5.M324 2010
    338.9'27—dc22   2009000467

Facts On File books are available at special discounts when purchased in bulk quantities for businesses, associations, institutions, or sales promotions. Please call our Special Sales Department in New York at (212) 967-8800 or (800) 322-8755.

You can find Facts On File on the World Wide Web at http://www.factsonfile.com

Text design by James Scotto-Lavino
Illustrations by Bobbi McCutcheon
Photo research by Elizabeth H. Oakes

Printed in the United States of America

Bang Hermitage 10 9 8 7 6 5 4 3 2 1

This book is printed on acid-free paper.

Contents

Preface

The first Earth Day took place on April 22, 1970, and occurred mainly because a handful of farsighted people understood the damage being inflicted daily on the environment. They understood also that natural resources do not last forever. An increasing rate of environmental disasters, hazardous waste spills, and wholesale destruction of forests, clean water, and other resources convinced Earth Day's founders that saving the environment would require a determined effort from scientists and nonscientists alike. Environmental science thus traces its birth to the early 1970s.

Environmental scientists at first had a hard time convincing the world of oncoming calamity. Small daily changes to the environment are more difficult to see than single explosive events. As it happened, the environment was being assaulted by both small damages and huge disasters. The public and its leaders could not ignore festering waste dumps, illnesses caused by pollution, or stretches of land no longer able to sustain life. Environmental laws began to take shape in the decade following the first Earth Day. With them, environmental science grew from a curiosity to a specialty taught in hundreds of universities.

The condition of the environment is constantly changing, but almost all scientists now agree it is not changing for the good. They agree on one other thing as well: Human activities are the major reason for the incredible harm dealt to the environment in the last 100 years. Some of these changes cannot be reversed. Environmental scientists therefore split their energies in addressing three aspects of ecology: cleaning up the damage already done to the earth, changing current uses of natural resources, and developing new technologies to conserve Earth's remaining natural resources. These objectives are part of the green movement. When new technologies are invented to fulfill the objectives, they can collectively be called green technology. Green Technology is a multivolume set that explores new methods for repairing and restoring the environment. The set covers a broad range of subjects as indicated by the following titles of each book:

- *Cleaning Up the Environment*
- *Waste Treatment*
- *Biodiversity*
- *Conservation*
- *Pollution*
- *Sustainability*
- *Environmental Engineering*
- *Renewable Energy*

Each volume gives brief historical background on the subject and current technologies. New technologies in environmental science are the focus of the remainder of each volume. Some green technologies are more theoretical than real, and their use is far in the future. Other green technologies have moved into the mainstream of life in this country. Recycling, alternative energies, energy-efficient buildings, and biotechnology are examples of green technologies in use today.

This set of books does not ignore the importance of local efforts by ordinary citizens to preserve the environment. It explains also the role played by large international organizations in getting different countries and cultures to find common ground for using natural resources. Green Technology is therefore part science and part social study. As a biologist, I am encouraged by the innovative science that is directed toward rescuing the environment from further damage. One goal of this set is to explain the scientific opportunities available for students in environmental studies. I am also encouraged by the dedication of environmental organizations, but I recognize the challenges that must still be overcome to halt further destruction of the environment. Readers of this book will also identify many challenges of technology and within society for preserving Earth. Perhaps this book will give students inspiration to put their unique talents toward cleaning up the environment.

Acknowledgments

I would like to thank a group of people who made this book possible. Appreciation goes to Bobbi McCutcheon, who helped turn my ideas into clear, straightforward illustrations, and Elizabeth Oakes, for providing wonderful photographs that recount the story of environmental medicine. My thanks also go to Marilyn Makepeace, Jacqueline Ladrech, and Jodie Rhodes for their tireless encouragement and support. I thank Melanie Piazza, director of animal care, and the staff at WildCare, San Rafael, California, for information on animal rehabilitation. Finally, I thank Frank Darmstadt, executive editor, and the editorial staff at Facts On File.

Introduction

The early 21st century may someday be looked upon as a pivotal point in the Earth's history. At its most dramatic, this era might someday be thought of as "the beginning of the end." The present decade marks a milestone in which the majority of people in industrialized nations and a large portion of people in the developing world are now feeling the effects of too many humans, too much waste, and the disappearance of plant and animal species faster than at any other time in history. The generations that will make up the first 100 years of this millennium may well determine whether the environment continues in a downward spiral or if technologies will emerge to change the way people need and use natural resources. The world truly seems poised to go either way.

Former U.S. vice president Al Gore became one of the first politicians in high national office to ask that the environment, particularly climate change, be made a priority. It may seem startling to realize that Gore made this request, not in the 1960s, when Rachel Carson's book *Silent Spring* opened the public's eyes to environmental pollution; it did not occur in the next decade with the first Earth Day in April 1970. Al Gore asked the world for a commitment to the faltering environment during the 1992 presidential race, but even after a half century of increasing evidence of environmental decay, many other leaders treated the environment as an afterthought. President George H. W. Bush went as far as to mock Gore, calling him "Ozone Man," and others continued to dismiss Gore's concerns in the following years. Whether Americans like it or not, the future well-being of the environment has a strong connection with politics; different administrations take different approaches to protecting natural resources while providing healthy conditions for business.

Though Vice President Al Gore received much criticism for insisting the environment become a political priority, U.S. leaders have included the environment in their political platforms for more than a decade now. In

July 2008 Gore challenged all political candidates to set definite goals for cleaning up the environment in an initiative that would become known as Repower America. In a climate conference in Poland, Gore said, "I ask you to join with me to call on every candidate, at every level, to accept this challenge—for America to be running on 100 percent zero-carbon electricity in 10 years. It's time to move beyond empty rhetoric. We need to act now." By the 2008 U.S. presidential election, the environment had become a leading issue, at least for one candidate.

In Senator Barack Obama's acceptance speech as Democratic Party candidate for the 2008 presidential election, he promised, ". . . as president, I will tap our natural gas reserves, invest in clean coal technology, and find ways to safely harness nuclear power. I'll help our auto companies retool, so that the fuel-efficient cars of the future are built right here in America." This was one of the few times in U.S. politics that the environment-business connection became a priority on a political platform.

Today politicians and large corporations figuratively wring their hands over the plight of the environment, a bit late perhaps, and surely some of these leaders embrace environmentalism only to win votes or satisfy stockholders. Whatever their motivation, the environment needs the help of government and industry to support the tireless work that local groups have been carrying out since the 1960s. Today's environmental situation is heading toward a type of critical mass, like pushing and pushing on a boulder until its weight shifts and it plummets downhill. Author Malcolm Gladwell described this phenomenon in 2000 as a "tipping point." Regardless of what this action is called, the planet is at a critical place. Either Earth's destruction may gain sufficient momentum to become impossible to stop or the world's environmentalists and leaders may gather enough support to turn back the destruction and change the way people care for the Earth.

Sustainability describes the innovations that will likely play a role in the near future for creating a critical mass in the environment's favor. The book opens with a chapter describing the ecosystem of humans, animals, plants, and other life. This chapter explains the concept of *ecological footprint* and the current and future impact of people covering the planet's surface. The chapter also investigates important points in environmentalism's history, including the first Earth Day. It also describes present-day perils in the environment, such as the crucial concept of *carrying capacity.* Finally, chapter 1 gives an overview of *deep ecology* and the philosophy of living for the environment rather than taking from the environment.

Chapter 2 offers a detailed look at the growing field of green biotechnology. It covers the new types of environmentalism in action today, the use of novel microorganisms to substitute for chemicals, nanotechnology, and the promise and the concerns that surround *bioengineering.*

Chapter 3 builds on the latest breakthroughs in biotechnology and other sciences to investigate new ways of producing food through sustainable agriculture. At least 40 years ago, scientists assured the public that the world food crisis would disappear as new forms of marine foods and microorganism-produced foods would come into use. That promise has not been fulfilled, and hunger is at crisis levels in a growing portion of the world. Sustainable agriculture has the weighty task of feeding hundreds of millions of people while it sustains the environment. If large agriculture simply decimated the environment in a rush to produce cheap food, the environment would soon be useless for everyone.

Chapter 4 presents the concept of white biotechnology. While green biotechnology tackles specific problems for improving the environment, white biotechnology incorporates other sciences and industry into its plans. White biotechnology also depends on the cooperation of governments working together to find new methods of creating sustainable lifestyles. For example, white biotechnology uses innovative chemistry and biology to invent materials that give benefits to people and the environment at the same time.

The next chapter discusses marine biotechnology because of the vital role played by the oceans in maintaining the Earth's life-sustaining conditions. This chapter describes current technologies for monitoring ocean *habitats* as well as the advances in plant and animal *aquaculture.* Finally, chapter 5 describes emerging plans for altering the ocean for the purpose of reducing global warming.

Chapter 6 looks at the applications of all these technologies by examining new materials that conserve natural resources. Many alternative woods, plastics, and products made from new *polymers* have already entered today's market. This chapter describes the benefits and some disadvantages of these materials and examines new inventions on the horizon.

Chapter 7 describes how to combine all of the technologies, new materials, public programs, and government programs to build sustainable communities. These communities might become the only hope for halting environmental loss, so they must be started in the near future rather than in some faraway time. This chapter highlights some places in the world that have begun making changes to live in a sustainable way. The chapter explains the planning that makes sustainable communities successful, and it also points out a few ambitious plans that did not work and why.

On October 12, 2007, Al Gore shared the Nobel Peace Prize with the United Nations Intergovernmental Panel on Climate Change (IPCC) for their efforts to alert the world to global warming. Gore said upon accepting the award, "The climate crisis is not a political issue, it is a moral and spiritual challenge to all of humanity." The same thing might be said about green technologies for sustainability.

Ecosystem Health

An *ecosystem* is a community of species that interact with one another and with their physical surroundings. In ecosystems, energy transfers from species to species in the form of food or prey, and this energy transfer works best when the ecosystem's members are all present and healthy. Ecosystems can be difficult to recognize at times because they range from very small systems to huge systems. For example, a tide pool no more than a few feet across represents an ecosystem; a coastline that stretches for miles also represents an ecosystem. In the case of the tide pool ecosystem, the tide pool also serves as a habitat for the invertebrates and vertebrates living there. A coastline holds many habitats: tide pools, rocks, sand, dunes, and marshes.

The Earth has always provided a way to support ecosystem health so that one member does not overgrow the system and dominate it and at the same time other members are able to persist, even if only in very small numbers. This natural balance is particularly important in what may be called *fragile ecosystems,* in which the system holds few species or it occupies a habitat that is easily destroyed. A sand dune is a fragile ecosystem because it contains few species that must depend on each other for energy and other services, especially compared with a forest ecosystem, which is complex and contains many species.

People affect ecosystem health in ways that are both subtle and obvious. A small pond ecosystem may lie in a woodland less than 50 yards from an interstate highway. Subtle influences such as traffic noise, sound vibrations from engines, increased heat near the freeway, and fumes all affect the activities of the species in the pond. Of course, obvious effects caused by humans include oil spills, trash, and filling in the pond for highway expansion. Even people with a love of nature upset ecosystems by their mere presence. The elk in Yellowstone National Park that stand in front of dozens of clicking cameras behave differently than elk living in remote parts of British Columbia, Canada. Author Myra Shackley wrote in 1996 in the book *Wildlife Tourism,* "Animals usually react to such encounters by trying to get away, which may cause severe exertion or displacement from home territory." Shackley also advised that studying the effects of people on wildlife can be difficult due to the unscripted actions of people and the different responses taken by very diverse wildlife.

This chapter discusses the growth of environmentalism when a handful of visionary people began to realize the effects of human activities on ecosystems. It discusses how the capacity of the land to sustain both natural ecosystems and humans has limits and the choices people have for saving an environment that has reached those limits. The chapter gives overviews of environmental technologies, lifestyles, and politics. It also stresses how environmentalism will probably never reach a time when environmentalists can rest, secure in the knowledge that everyone will do the right thing for their environment.

HISTORY OF ENVIRONMENTALISM

Environmental science is a field of study that draws on many disciplines to learn how the Earth and its living things work. Environmentalism is not a science but a political or social movement that works to improve the environment as well as all the planet's *biota,* or living things. Over the past century, environmental scientists have discovered facts about the Earth that have given environmentalism new areas of attention. For example, *Silent Spring*'s author, Rachel Carson, was more an amateur scientist than a trained environmental scientist—environmental science did not even exist when Carson wrote her groundbreaking books—but her theories on pesticides aroused environmentalists to confront the dangers of pesticides on people and animals. At other times, environmentalism took the lead on issues that prompted environmental science's new technologies. In the early 1990s, for example, consumers increasingly questioned the merit of using paper or plastic grocery bags; many shoppers began carrying reusable bags for their groceries. The plastics industry soon responded by making bags out of recyclable plastic. Without environmentalism, would industry ever have achieved this simple improvement?

Environmentalism most likely began with the European explorers who traveled through North America, the jungles of South America, and the polar regions. Each team of explorers had individuals who recorded their thoughts and unwittingly laid the foundation for both environmentalism and environmental science. Botanical illustrator William Bartram became one of the first such environmental historians by illustrating his trips through the southeastern states. In one visit to Florida's St. Johns River area in 1791, Bartram wrote of his disquieting observations: "At about fifty yards distance from the landing place, stands a magnificent Indian mount. About fifteen years ago I visited this place, at which time there were no settlements of white people, but all appeared wild and savage; yet in that uncultivated state it expressed an almost inexpressible air of grandeur, which was now entirely changed. At that time there was a very considerable extent of old fields round about the mount; there was also a large orange grove, together with palms and live oaks, extending from near the mount, along the banks, downwards, all of which has since been cleared away to make room for planting ground." Bartram's comment foreshadowed the environmental damage to come.

The American leaders of Bartram's day, Thomas Jefferson, George Washington, and James Madison, also harbored a desire to preserve the land. Each of these U.S. presidents owned agricultural land, and they tried to make their lands more productive while at the same time sustaining the soil, water, and natural growth for their grandchildren. At this time when the United States was making attempts at prudent land management, Europe had already hit a population explosion, inspiring essayist Thomas Malthus to publish a warning of the coming disaster from too many people, too little food, and few safe places to live. Malthus's 1798 "Essay on the Principle of Population" raised modest interest in Europe. Americans, however, largely ignored the essay because the United States stretched for thousands of miles, and people likely felt their future to be equally as limitless.

At the dawn of the 19th century, the vast spaces west of the Mississippi River became an ambitious experiment in environmental science. Meriwether Lewis and William Clark recorded hundreds of new species of plants and animals as they led their Corps of Discovery from the Missouri River to Oregon from 1804 to 1806, under the decree of President Thomas Jefferson. In his notes, Lewis showed he was an environmentalist at heart as he wrote of the unspoiled lands the group traveled through. At the same time, lesser-known naturalist Alexander von Humboldt traveled the west coasts of North and South America, developing theories on the relationships between the land, its biota, and the humans who shared it. While Lewis and Clark had put forth the first detailed observations of this continent's natural world, von Humboldt was inventing the concepts of ecology and the ecosystem.

The U.S. population began rapid expansion in the mid-1800s, just as Europe had a century earlier. The northern states grew more industrialized—and dirty with pollution—with the start of the Civil War, and city dwellers probably disliked the state of their cities. These cities contained rivers that carried dumped garbage, questionable sewer systems, and poor sanitation. During this time, Charles Darwin and Gregor Mendel proposed theories on how nature evolved; John Muir, Henry David Thoreau, and George Perkins Marsh wrote of the value of pristine mountains and water, not yet destroyed by human intrusion.

President Theodore Roosevelt made the importance of natural resources a primary part of his two terms (1901–09) by focusing on forests and wildlife. After his presidency, Roosevelt continued his devotion to the environment by exploring Brazil's Amazon region. By the time Roosevelt died in 1919, however, U.S. cities were expanding fast, and industries were gobbling up all resources possible to supply the wave of industrialization. Wildlife sportsmen may have been the first to notice that natural resources were vanishing. The loss of undisturbed forests, rivers, lakes, and coasts meant the disappearance of wildlife. In 1925 George Grinnell and Charles Sheldon of the Boone and Crockett Sportsmen's Club wrote, "The original purpose of the Boone and Crockett Club, to make hunting easier and more successful, has changed with changing conditions, so that now it is devoted chiefly to setting better standards in conservation." The environmental movement took shape from that point forward. Landmarks of today's environmentalism are summarized in the following table.

The first environmentalists endured criticism and derision from politicians, industry, and the public. Even today the term *tree hugger* is meant to insult environmentalists rather than recognize their efforts to preserve the planet for future generations. Today politicians take note of the environment in their speeches, and schools teach young students ways to curb natural-resource overuse. Local governments pitch in on each Earth Day celebration, discussed in the sidebar "Earth Day." This new, wide-ranging

LANDMARKS IN THE HISTORY OF ENVIRONMENTALISM		
EVENT	**YEAR**	**SIGNIFICANCE**
Thomas Malthus's "Essay on the Principle of Population" is published	1798	publically decried the potential problems caused by population growth
John James Audubon's *Labrador Journals* is published	1840	recognizes the wholesale destruction of natural resources in North America
Henry David Thoreau's writings are composed	1845–64	describe the value of all living things in nature
Theodore Roosevelt is in office	1907	address to Congress on conservation of natural resources
Henry Beston publishes *The Outermost House*	1928	book explores man's relationship with nature and serves as inspiration for Rachel Carson's writings
Rachel Carson publishes *Silent Spring*	1962	signaled the beginning of the public's concern about pollution
first Earth Day (see sidebar)	1970	symbolic beginning of modern environmentalism
Arne Naess proposes deep ecology	1973	relates economic and social needs to environmental needs
Greenpeace's Declaration of Interdependence is issued	1976	stated unequivocally that humans are leading the destruction of the Earth
Al Gore publishes *An Inconvenient Truth*	2006	alerts the public to the environment's rapid decline

scope of environmentalism has not come a moment too soon, because the Earth has sustained some serious injuries in the past century since the Industrial Revolution.

THE HEALTH OF OUR BIOSPHERE

The term *biosphere* refers to the part of the Earth containing life. The biosphere encompasses the lower atmosphere, called the troposphere, plus the planet's surface, deep soils, and deep ocean. The Worldwatch Institute, based in Washington, D.C., is an organization that has taken the lead in measuring the health of the biosphere, and in fact, this organization produces a yearly summary titled *Vital Signs* that reports on aspects of the environment: population growth, globalization of economies, climate change, vehicle production, trends in using alternative energy sources, fish harvests, grain output, fossil fuel use, deforestation, and *biodiversity* loss. Any or all of these measures have been used to assess the state of the planet, but population growth, climate change, and biodiversity loss may be the strongest indicators of the biosphere's health. This is because these three interrelated subjects connect with many other trends in the environment. For example, biodiversity loss is usually an overall indication of overfishing, large-scale agriculture, or deforestation.

Some parts of the world have improved their environment. Air pollution laws have cleaned up the atmosphere in many parts of the United States; forested land has expanded in Europe; and the solar, wind, biofuel, and hydropower industries are gaining ground rapidly. Overall, however, Earth's environment continues to change in troubling ways. Carbon dioxide (CO_2) emissions receive scrutiny as an environmental ill because the levels of this *greenhouse gas* in the atmosphere indicate large increases in population, vehicle use, industry, and deforestation. The most dramatic effect of rising CO_2 levels relates to climate change, specifically global warming. Global warming is the increase in the average temperature of the Earth's atmosphere due to increased greenhouse gases caused by human activities.

The world's temperatures have not stayed within a small range throughout history. Instead, average temperatures in the troposphere fluctuate as much as 41–43°F (5–6°C) from century to century. This fluctuation has made some members of the public and even a few atmospheric scientists question whether the climate is truly changing in an unnatural

Carbon Dioxide in the Atmosphere

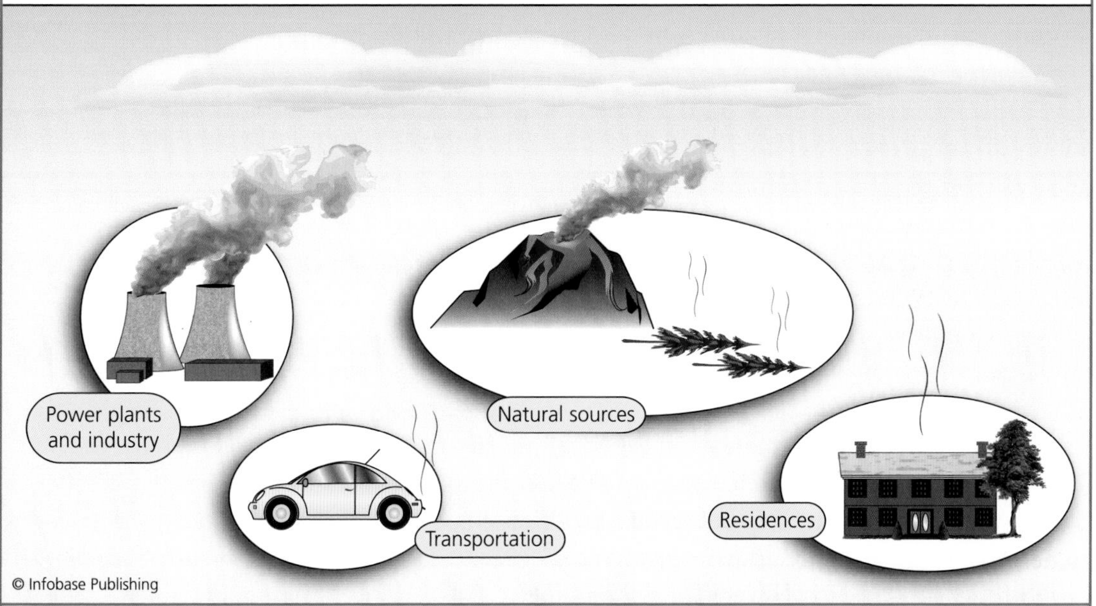

Power plants and industry

Natural sources

Transportation

Residences

© Infobase Publishing

Carbon is the main constituent in all biota. Carbon cycling through the Earth, biota, and the atmosphere affects energy transfer from the Sun to living things. Human activities have caused an imbalance in the Earth's natural carbon cycle by producing excess amounts of CO_2, which contribute to global warming.

way. Scientists who question global warming make headlines, but actually they make up a very small portion of the large scientific community that has collected overwhelming evidence that human activities cause temperatures to increase and have done so since the Industrial Revolution.

The public becomes less certain of science when news stories on climate change seem to contain as many doubters as believers. Two factors have led to the misperception that climate change has not been proven: (1) the news media always seeking opposing opinions on topics related to the environment and (2) scientists viewing the world as containing very few things that are 100 percent certain. The hallmark of good science resides in scientific challenges to theory. The public may fail to understand that differing opinions make up any scientific discourse, and people may therefore conclude that scientists disagree on issues such as global warming.

Scientific opinion often loses some of its meaning between the laboratory and a news story. Risk analysis expert Kimberly Thompson, a professor at Harvard University, warned in the *New York Times* in 2008, "Words that we as scientists use to express uncertainty routinely get dropped out to make [news] stories have more punch and be stronger." Thompson explained that terms such as "results suggest" or "it is likely that" must be retained when reporters write about science because "they convey meaning to readers not only in the story at hand, but more gener-

EARTH DAY

arth Day is an annual global event that has come to represent two important features of environmentalism. First, Earth Day symbolized the awakening of the public to damage being inflicted on the environment and a willingness to do something about it. Prior to the first few Earth Days, only a small percentage of the public plus a limited number of scientists put much thought into environmental decay. Second, Earth Day emphasized the global nature of preserving the environment by drawing upon governments, private organizations, and industries to work with citizens and scientists on environmental issues.

Earth Day is a daylong recognition of the Earth's ecology, and it also serves to explore new technologies for preserving natural resources. Residents of San Francisco, California, proclaimed the first Earth Day in 1970 as a teach-in on the environment, modeled on similar gatherings focused on the Vietnam War (1959–75), to be held on or near April 22 each year. That first celebration drew close to 20 million people worldwide. The first few Earth Days raised the consciousness of more and more people about care of the environment.

With each passing year, Earth Day has focused on specific aspects of ecology. In the decades since the first Earth Day, succeeding celebrations have addressed the following topics: preservation of rain forests; waste reduction; banning the logging of ancient forests; recycling and composting; acid rain prevention; and slowing the production of greenhouse gases. In the year 2000 the Earth Day Network launched Earth Month to draw global participation in environmental activities. Earth Month April 2000 involved an estimated one-third of the world's population to address a variety of issues, especially climate change and pollution. (The actual Earth Day in 2000 attracted 500 million people worldwide.)

On 2008's Earth Day in Washington, D.C., the Earth Day Network's president, Kathleen Rogers, said to the *Washington Post*, "This is the entry point for people to help with environmental change. We need to engage everybody in the fight against global warming, and we need to get Congress to know that what it is doing is not enough." Rogers made two important points. First, Earth Day's outreach to regular citizens has been invaluable in educating everyone on environmental issues. Sec-

ally about science being less precise than is typically conveyed." In fact, translating scientific opinion into public opinion is one of the biggest hurdles in teaching the public about sustainability.

Judging by the three main parameters—population growth, climate change, and biodiversity loss—the current condition of the biosphere's health is not good. Human population growth rates have declined, but

ond, countries need more progress in getting government and industry leaders to commit to long-term environmental programs. The 2000 Earth Day celebrations created the six following objectives for subsequent Earth Days to ensure the biosphere's health improves rather than declines:

1. Empower all citizens to face environmental challenges worldwide
2. Create global networks for organizing major programs
3. Serve as a communication resource for groups with the same objectives
4. Highlight innovative technologies
5. Pressure national leaders to pursue clean, renewable energy
6. Inspire cultural shifts toward environmental care

The *New York Times* reporter Gladwin Hill described the preparations for that first Earth Day in 1970: "Thousands of colleges, schools and communities across the country were getting ready yesterday for an unprecedented event: tomorrow's Earth Day—an interlude of national contemplation of problems and man's deteriorating environment." Interesting, too, was the fact that the early participants had not yet realized the power of teaming with government to build a stronger coalition. Hill wrote, "At least several dozen members of Congress and a number of federal officials will be participating in Earth Day activities all over the country, although there is little or no federal involvement. Teach-in leaders, wary of such involvements, lest it appear that the movement has been 'captured' by the Nixon Administration, said they had turned down a White House invitation in recent weeks for a discussion session because 'we didn't feel there was a great deal to chat about.'" Times have changed, and so has Earth Day. Though animosity exists at times among environmentalists, government, and industry, these three groups have made extraordinary progress in communicating their desires to one another. Earth Day often serves as a symbolic starting point for their discussions.

total population growth continues upward. In 2005 the world's population (6.45 billion) had doubled its 1950 figure; the population may exceed 9 billion people by 2050. Climate change has accelerated in the past 200 years to the point where glaciers have begun to melt, sea levels are rising, and vegetation and animals are undergoing stresses due to warmer temperatures. Biodiversity loss has also approached crisis levels. The biodiversity scholar Edward O. Wilson has taken on the difficult task of estimating biodiversity loss. In Wilson's classic 1988 book on the subject, *Biodiversity*, he estimated 17,500 species were lost every year. "Given 10 million species in the flora and fauna of all the inhabitants of the world, the loss is roughly one out of every thousand species per year." This figure adds up to very alarming losses in a person's lifetime.

HOW ECOSYSTEMS WORK

Ecology is the science of relationships between living organisms and their environment. Ecosystem health is a major part of ecology because ecosystems encompass all of the living and nonliving things in a particular environment and their relationships. Humans who lived as hunter-gatherers more than 10,000 years ago fit into ecosystems by acting as predators, and on occasion they probably also served as prey. These early humans also picked seed-bearing fruits and eventually dispersed the seeds to help propagate new plant growth. By scratching through the soil to look for root vegetables, people helped aerate the soil for other plants.

The role of humans in today's ecosystems differs from that of early human settlements. Today, humans in almost every part of the world, except for a small number of remote primitive tribes, no longer interact with nature as they once did. People have detached themselves from most ecosystems. Modern humans also tend to negatively affect ecosystems in ways that the earliest civilizations did not. These negative effects are mainly the result of population growth and industry.

One aspect of ecosystem study in environmental science provides a clue as to how disconnected humans have become from nature. Many ecosystems are named for the dominant species within them. Therefore, the world contains coral reef ecosystems, evergreen forest ecosystems, grassland ecosystems, and so on, but environmental science contains no "human ecosystems."

Ecosystems exist in every place on Earth. Most ecosystems contain food webs of varying complexity that make up one type of natural capital: biodiversity. This tidal stream feeding the Chesapeake Bay possesses interrelationships between aquatic species, animal life that lives or feeds in the mud flats, and the banks' vegetation. Actions that sustain the Earth's freshwater, marine waters, coastlines, and species also sustain thousands of ecosystems like this one. *(April Bahen; NOAA's Estuarine Research Reserve Collection)*

Ecosystems at their best are those that contain a proper balance of members, each serving a beneficial role. A balanced ecosystem contains proportions of different species so that no one species dominates all the others and drives them to extinction. Most ecosystems contain a foundation of plant species that capture the Sun's energy plus diverse microorganisms, also called *microbes,* and tiny invertebrates upon which small animals feed. An ecosystem usually contains *food chains* of increasingly larger predators, with a small population of the largest predators at the highest level. Complex ecosystems additionally contain *food webs* in which various food chains interconnect. An imbalanced ecosystem may result from the following events: invasive species that overwhelm a habitat or ecosystem; disease; habitat loss; climate change; or over-predation. Humans, for instance, have been the cause of over-predation in certain marine waters where overfishing has depleted fish breeding grounds and removed almost all fish from an area of the ocean. As a result, these over-fished areas have severely damaged ecosystems.

Can people live normal lifestyles without damaging ecosystems? It seems only ecosystems far from civilization enjoy minimal effects from human activities, and even these remote places have probably been touched by air or water pollution. In order for people to preserve the workings of ecosystems, they must behave in a way that maintains *ecosystem stability,* meaning a condition in which an ecosystem can rebound from events causing it temporary damage. This stability can be measured in any of the three following ways:

1. Variation—Ecosystems that fluctuate only slightly during changing conditions are more stable than ecosystems that undergo great fluctuations.
2. Resistance—Ecosystems containing species that resist disturbances and maintain their population size are more stable than ecosystems containing species sensitive to disturbance.
3. Resilience—Ecosystems containing species that quickly return to normal population size after a disturbance are more stable than ecosystems containing species that cannot recover quickly.

The more interconnected species are within an ecosystem, the better the chances of the entire ecosystem's survival. Since an ecosystem may be thought of as the central unit in all of the Earth's ecology, humankind's greatest contribution to the environment is to protect ecosystem health. This goal is far easier to talk about than it is to achieve; people have already overrun many ecosystems by their sheer numbers and urban expansion into previously undisturbed places.

CARRYING CAPACITY

Carrying capacity refers to the maximum population size of a species that a habitat or ecosystem can support over time without degrading the environment. Habitat has been decreasing worldwide for thousands

Sustainability is the capacity of a system, such as the human population, to survive for a finite period of time. Overpopulation in some parts of the world dramatically affects sustainability because dense populations strain water, air, fuel, and food resources and produce large amounts of waste. Bangkok, Thailand, shown here, is one of many cities with a rapidly growing population. In less than 40 years, Bangkok's population has increased fourfold. *(Jan and Sylvio)*

of plant and animal species for a variety of reasons, mainly habitat destruction, pollution, and fragmentation. In habitat fragmentation, structures such as highways cleave a habitat into smaller pieces, which then cannot sustain a healthy population of the species living there. Many habitats and ecosystems may be reaching or have already reached their carrying capacity. In this situation, competition between species and within species for space, food, water, and shelter reaches critical levels.

Natural populations rarely exceed their carrying capacity because species increase or decrease breeding or the number of offspring to meet available food supplies; other species migrate every few generations to new habitat. Humans now compete against animals and plant life for space, water, and food, and human communities often fragment habitats, which hastens extinction rates. Natural extinction rates vary from about 10 to 100 species per year; today's accelerated rate is 27,000.

Extinction affects people in subtle ways. The Evolution Library of the Public Broadcasting System (PBS) has explained the problem of ecosystem and biodiversity loss: "For along with that [loss of] species richness, the ecosystem is likely to lose much of its ability to provide many of the valuable services that we take for granted, from cleaning and recirculating air and water, to pollinating crops and providing a source for new pharmaceuticals. And while the fossil record tells us that biodiversity has always recovered, it also tells us that the recovery will be unbearably slow in human terms—5 to 10 million years after the mass extinctions of the past. That's more than 200,000 generations of human-kind before levels of biodiversity comparable to those we inherited might be restored." Ecosystems thus interweave with human life, nonhuman life, and the Earth's natural rhythms.

Animals that reach their habitat's carrying capacity may respond in three main ways: migrate to a new habitat, alter their diet, or produce less offspring. Humans respond to carrying capacity in a very different

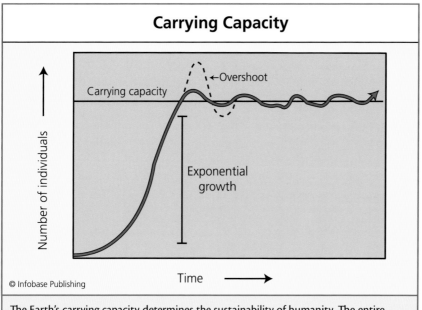

Carrying Capacity

Number of individuals

Carrying capacity

←Overshoot

Exponential growth

Time

© Infobase Publishing

The Earth's carrying capacity determines the sustainability of humanity. The entire population has entered an overshoot situation in which humans use more resources and produce more wastes than the Earth can support. No population can increase in size indefinitely. Green technologies can prolong sustainability, but not forever.

way. Technologies in food production, trade, energy generation, and the supply of consumer products have enabled many people to live in densely populated habitats that would not normally sustain such a large population. The Earth's human carrying capacity has not yet been determined, but some scientists fear the world is coming dangerously close to reaching it. A Cornell University professor of agricultural sciences, David Pimentel, was quoted in the college newsletter in 2004 pointing out, "If we refuse to reduce our numbers ourselves, nature will find much less pleasant ways to control human population: malnourishment, starvation, disease, stress and violence." Using natural resources sustainably helps prolong human life into the future.

Talking about sustainability has become commonplace in environmental discussions, so much so perhaps that it begins to lose its meaning. Sustainability is critical, because as people accept more sustainable lifestyles, they postpone the possibility of reaching the Earth's human carrying capacity.

INDICATORS OF ECOSYSTEM HEALTH

The impact of human activities on the environment relates directly to ecosystem health. Since 1992 the human effect on the environment has been expressed by the ecological footprint, which is a calculation of how much water and land a population needs to produce the resources it consumes and degrade the wastes it produces. The ecological footprint relates to sustainability in that it gives an indication of whether current lifestyles can continue into the future.

The Global Footprint Network estimates that the world's human population has exceeded its ecological footprint by 23 percent, meaning people are using up natural resources and creating wastes faster than the Earth can replace them or degrade them, respectively. To put it simply: people are using up nature. In 2003 the ecological footprint had exceeded 25 percent, but it has since declined slightly due to aggressive conservation and antipollution programs. When any population exceeds its carrying capacity, it enters a condition called *ecological overshoot* in which the population must deplete natural resources just to sustain its current rate of growth.

Several components go into an ecological footprint calculation. Though the world has exceeded its available natural resources, each country differs: some countries have greatly exceeded their ecological footprints, and other countries have stayed well within sustainable levels. These differences arise from several factors, namely, the proportion of developed land in a country, its agriculture lands, fishing grounds, and waste production. These factors together represent *consumption land use.* The following factors make up a country's consumption land use:

- developed land cropland
- grazing land
- fishing ground
- forest
- CO_2-sequestering land (the amount of land that absorbs carbon dioxide produced by a specific region)

By calculating consumption land use, cities, states, countries, and entire continents can identify the activities that consume the most natural resources and affect the ecological footprint. In the worst case, a region has exceeded its biological carrying capacity, called biocapacity. Biocapacity is the ability of the land's ecosystems to support its human population. Canada's Yukon Territory, for example, has not exceeded its biocapacity, but Tokyo, Japan—the world's most populous city—has exceeded its biocapacity, probably by a wide margin. A land's biocapacity derives from multiplying land area by the land's yield factor—the productivity of the land—and an equivalence factor that standardizes different types of land (urban, cultivated, rangeland, etc.) to a unit of area called the *global hectare.*

biocapacity = land area (hectares) × yield factor × equivalence factor

These calculations use hectares; a hectare is a unit of land area equaling 2.47 acres. Like biocapacity calculations, the ecological footprint calculations use global hectares.

The goal of any region interested in conserving resources is to create either a *biocapacity buffer* or an *ecological reserve,* or both. A biocapacity buffer refers to land set aside to maintain healthy ecosystems and species. A wilderness area, for instance, represents a type of biocapacity buffer. An ecological reserve results when a population's ecological footprint does not exceed the land's biocapacity—it leaves something in

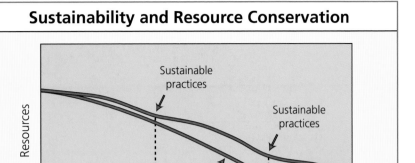

Sustainability and Resource Conservation

Sustainable practices

Sustainable practices

No sustainable practices

Resources

Time ⟶

© Infobase Publishing

Activities that conserve resources have a direct positive effect on prolonging sustainability. This graph illustrates the need for new technologies in the near and distant future. Over time, even alternative materials that are invented to conserve natural resources will become scarce. Technology will therefore always be critical for supporting life on Earth.

reserve. By contrast, an *ecological deficit* means a population's ecological footprint has exceeded its biocapacity (e.g., Tokyo). The following table lists countries with the greatest reserves and countries with the largest deficits.

High-income countries use about 6.4 global hectares (16 global acres) of biocapacity per person, while middle-income countries use 1.9 global hectares (4.6 global acres) per person, and low-income countries use only 0.8 global hectares (2 global acres) per person. The global average of bio-capacity in global hectares per person is about 2.2 (5.4 global acres). The United States currently uses 9.6 global hectares (23.7 global acres) of its biocapacity per person.

The data presented here may soon change from larger to smaller deficits and from negative reserves to positive reserves if countries commit themselves to sustainable activities. Not all changes must be drastic lifestyle decisions; the public can manage small changes for big effects on the environment. But what if small changes in behavior equates to too little, too late? Members of the environmental movement have called for stronger measures, discussed in the "Deep Ecology" sidebar.

ECOLOGICAL FOOTPRINTS BY COUNTRY (GLOBAL HECTARES/PERSON)			
COUNTRY	RESERVE	COUNTRY	DEFICIT
Gabon	17.8	United Arab Emirates	-11.0
Bolivia	13.7	Kuwait	-7.0
New Zealand	9.0	United States of America	-4.8
Mongolia	8.7	Belgium and Luxembourg	-4.4
Brazil	7.8	Israel	-4.2
Canada	6.9	United Kingdom	-4.0
Uruguay	6.1	Saudi Arabia	-3.7
Australia	5.9	Netherlands, Spain, Greece	-3.6
Mauritania	4.5	Italy	-3.1
Finland	4.4	Germany	-2.8

Note: To convert to global acres, multiply by 2.47.
Source: Global Footprint Network

Other footprint measurements are also useful in evaluating the health of an ecosystem. The following table describes the main indicators of a healthy environment.

TECHNOLOGIES FOR ECOSYSTEM STUDY

Ecosystem studies focus on how specific ecosystems work and consist of the following specialties:

- energy and matter transformations
- ecosystem composition and structure
- ecosystem dynamics (changes in an ecosystem over time)

ECOLOGICAL INDICATORS IN ADDITION TO ECOLOGICAL FOOTPRINT	
INDICATOR	**DESCRIPTION**
national footprint	ecological footprint of a country
carbon footprint	amount of all greenhouse gases produced by a region
energy footprint	sum of all area used to produce nonfood and non-feed energy
consumption footprint	area used to produce materials a defined population consumes and to absorb the wastes it produces
per capita consumption footprint	area used to produce materials a person consumes and absorb the wastes produced
primary production footprint	sum of all areas to produce all harvestable products, to support all built structures, and to absorb all the fossil fuel emissions
nuclear footprint	ecological footprint of the electricity generated by nuclear power (estimated as 8 percent of carbon footprint)

- connections between certain ecosystems and other factors in the environment
- disturbed ecosystems
- *ecosystem modeling*

The areas of study listed here apply to the Earth's various ecosystems such as coastal, marine, forest, grassland, riparian, and polar. Ecosystem science includes studies on an ecosystem's biological, physical, and chemical features, and its main methods depend on monitoring and measuring. Monitoring refers to any technique for counting or otherwise keeping track of an ecosystem's components. For instance, a bush pilot may fly

DEEP ECOLOGY

Norwegian philosopher Arne Naess proposed the term *deep ecology* in 1973 to describe a new viewpoint on how humans must live, that is, by accepting the notion that all people are connected to something greater than themselves and human needs are not biology's primary needs. Before Naess gave a name to this theory, American naturalist and wildlife manager Aldo Leopold published *A Sand County Almanac* in 1949, which was a collection of essays on the moral principles of land use and sharing land with the rest of biota. Leopold had proposed what is now known as biocentric equality, meaning that all natural things have a right to exist. Simple enough, it would seem, yet many people reject biocentric thinking in favor of believing the world should be human-centered. Regardless of point of view, the main reason for biocapacity deficit is that humans satisfy their needs at the expense of the environment.

Professor Stephan Harding of Schumacher College in Britain expressed the essence of deep ecology in an online article "What Is Deep Ecology?" Harding said, "As a wildlife manager of those times [1920s], Leopold adhered to the unquestioning belief that humans were superior to the rest of nature, and were thus morally justified in manipulating it as much as was required in order to maximize human welfare. One morning, Leopold was out with some friends on a walk in the mountains. Being hunters, they carried their rifles with them, in case they got a chance to kill

over an African savannah to monitor the lion prides living in a region. Measuring involves different types of scientific analysis that determine the quantities of chemicals, minerals, gases, temperatures, rainfall, or animal or plant species. A pilot who actually counts the number of prides within a given area is taking a measurement.

Ecosystem science also uses sensitive techniques in molecular biology for studying the genetic makeup of plant and animal species. This discipline is known as *genomics*, meaning the study of all the genes of a single animal or species. To do this, scientists rely on polymerase chain reaction (PCR) technology, which takes a small piece of deoxyribonucleic acid (DNA) and multiplies it many times to create a much larger amount of DNA that scientists can then analyze, usually by two methods: DNA hybridization and DNA sequencing. DNA hybridization matches pieces of DNA from two different individuals to determine how closely they are related. Ecological studies that find only closely related individuals may

some wolves. It got around to lunch time and they sat down on a cliff overlooking a turbulent river. Soon they saw what appeared to be some deer fording the torrent, but they soon realized that it was a pack of wolves. They took up their rifles and began to shoot excitedly into the pack, but with little accuracy. Eventually an old wolf was down by the side of the river, and Leopold rushed down to gloat at her death. What met him was a fierce green fire dying in the wolf's eyes. He writes in a chapter entitled 'Thinking Like a Mountain' that: 'There was something new to me in those eyes, something known only to her and to the mountain. I thought that because fewer wolves meant more deer, that no wolves would mean hunter's paradise. But after seeing the green fire die, I sensed that neither the wolf nor the mountain agreed with such a view.'" This realization guided Leopold for the rest of his career in wildlife management.

Deep ecology represents a special way of thinking about the environment in which a person must consider the larger picture of life. Put another way, deep ecology requires self-realization. Deep ecology cannot in itself correct climate change, reverse biodiversity loss, or slow population growth, but this philosophy may in fact become the only way for people to understand that the planet does not belong to humans, but rather humans are connected to everything around them and every act eventually affects some part of the environment.

indicate unusual levels of inbreeding, perhaps due to habitat fragmentation or a declining population. DNA sequencing enables scientists to inspect the gene makeup of an individual or a species to learn one of three things: (1) the relatedness of species; (2) species groups with an adaptation that allows them to persist in an ecosystem; and (3) health strengths or weaknesses that influence the survival of ecosystem members.

Environmental scientists also study ecosystems using a combination of three technologies that work on a larger scale than DNA: (1) field studies, (2) remote studies, and (3) laboratory studies. Field studies have been the backbone of environmental science. They involve on-the-ground collection of samples, taking measurements, and counting species or individual plants or animals. Field studies combine sophisticated instruments with these on-the-ground activities. Electronic instruments measure ecosystem features such as nutrients, soil conditions, water constituents, weather conditions, and atmospheric components.

Scientists often combine field studies with remote studies, in which aircraft or satellites gather data on large swaths of continents or oceans. This technology makes use of geographic information systems (GIS) to help scientists map certain features of the environment. Remote sensing has been useful for tracking ocean pollution, monitoring coastline decay, monitoring forest loss and the health of forests, air emissions, and sea temperatures. Remote studies also enable environmental medical professionals to connect cancer cases with known pollution sources. Finally, laboratory studies support both field and remote studies in a study technique called ecosystem modeling.

Ecosystem models depict real-life conditions in miniature. Sausalito, California, is home to the U.S. Army Corps of Engineers, Bay Model, which is a 1.5-acre (0.6 ha) indoor automated model of San Francisco and San Pablo Bays. The three-dimensional model, when filled with water, has been used by ecologists and engineers to simulate the following bay conditions: tides, currents, sediment movement, fresh- and saltwater mixing, *saltwater intrusion,* pollution, and the effects of structures built in the bay. Computer programs sort data collected from sensors positioned in the model, while scientists add dyes to the water at specific points along the model's shoreline. The dye concentrations are measured

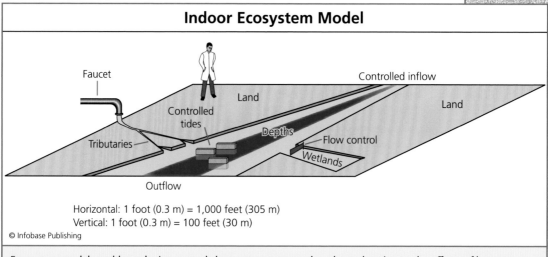

Indoor Ecosystem Model

Faucet

Controlled inflow

Land

Land

Controlled tides

Depths

Flow control

Tributaries

Wetlands

Outflow

Horizontal: 1 foot (0.3 m) = 1,000 feet (305 m)
Vertical: 1 foot (0.3 m) = 100 feet (30 m)

© Infobase Publishing

Ecosystem models enable ecologists to study how ecosystems work and test theories on the effects of human activities. This model of San Francisco Bay can create tides for studying the effects of storms, flooding, erosion, pollution, sediment flow, saltwater and freshwater mixing, and replenishment of wetlands. Future models may be designed to predict the depletion of other natural capital.

on a fluorometer, which analyzes samples by measuring wavelengths of fluorescent light given off by a compound. As the dye moves with the model's simulated currents, the fluorometric data help identify unique patterns in bay currents.

Models have become powerful tools in analyzing ecosystems because their programs can detect trends and relationships in large volumes of data. In turn, scientists may run hypothetical situations to gauge the damage that may be caused to an ecosystem due to natural events or human activities. Isolated laboratory studies would likely not simulate the exact conditions in the environment in every experiment. Therefore, any results from laboratory experiments must always be confirmed with additional field studies.

OUR COMMITMENT TO THE ENVIRONMENT

Various individuals certainly hold very different commitments to the environment, ranging from no interest at all to lifestyles such as deep ecology. The average citizen's commitment might not extend beyond such easy tasks as recycling, carpooling, and attending an Earth Day gathering from time to time.

Perhaps the environmentalists' greatest challenge is to stir society into an urgency regarding environmental decay. Former U. S. vice president Al Gore described society's commitment to the environment in his Oscar-winning 2006 documentary, *An Inconvenient Truth*. Gore referred to the "boiling frog syndrome," a metaphor for how people react to change. The boiling frog syndrome proposes that a frog dropped into boiling water will quickly jump out, but if dropped into cool water that is slowly raised to a boil, the frog will not flee, and in fact it will boil to death.

The boiling frog syndrome portrays how people react to change. Dramatic sudden changes instigate resistance and even anger in many individuals. Conversely, people do well at adjusting to slow change that occurs over a long period of time. Environmental changes have happened on a very large scale, and with the exception of isolated accidents, environmental change occurs in a gradual way. For example, air pollution may not seem much worse than it was the day before, so people may assume air pollution is not a serious problem. What has happened instead is that

People have a difficult time noticing destruction of the environment because the decay often takes place over a span of years. Events such as coral bleaching, melting of polar ice, or the disappearance of deciduous forests may not look different from one year to the next. Over a decade, however, differences can be striking. By monitoring glaciers, such as Austria's Pasterze glacier shown here, over many years, scientists have identified alarming changes. *(Susannah Sayler/The Canary Project)*

people have adjusted daily to a slow buildup of air pollutants. The world's people may have undergone a type of desensitization toward environmental decay that developed over a lifetime or longer.

Grassroots organizations, associations of ordinary citizens who band together to initiate change, have often been the sole reason that people worry about the environment and do not fall victim to desensitization. Grassroots groups gained momentum in the 1980s as more and more people realized they needed to make major changes in order to slow the degradation of their surroundings. By the 1990s entrepreneurs joined the cause by developing new technologies for cleaning up pollution, degrading wastes, and conserving resources. Some grassroots organizations continue to work on local environmental issues, but others have taken steps to build larger initiatives with governments. The green movement represents the main environmental-political coalition today.

The green movement has developed a strong voice in the public arena and in politics. It has encouraged leaders to address environmental worries and to seek commitment from industry to provide products that conserve natural resources. It is difficult to say when this shift from grassroots interest to political force occurred in the United States and other countries. In truth, many different events probably came together to make the green movement a legitimate part of society.

In the United States, the 1986 oil spill by the tanker *Exxon Valdez* in Alaska's Prince William Sound likely played a crucial role in strengthening this country's green movement. Similar environmental accidents had the same effect in Europe. But less dramatic events have contributed in their own ways to the green movement. A 2008 business news release stated, "Boulder [Colorado] is often referred to as the 'Birthplace of the Green Movement' because of locally based companies . . . who [sic] back in the 1970s, made a choice to advocate and run a business that would promote a healthier, more sustainable lifestyle. Not only did they make these eco-conscious decisions for themselves, but they were leading by example, encouraging others to consider the environment, society and the overall health and well-being of our planet. They didn't know it at the time, but they were changing the world." This describes the strength that comes from individuals working for a common cause.

The green movement will certainly continue at the local level, but its greatest effect has been in industry and politics. The following "Case Study: Green Politics and Elections" discusses the arrival of environmentalism in world politics.

GREEN LIFESTYLE CHOICES

Every person in the industrialized world is able to make small or large changes that help the environment. Television, the Internet, and the print media abound with tips for saving energy, reducing wastes, and helping restore damaged habitats. Each person can also take the initiative to decide if there are additional ways he or she might make a difference. The green lifestyle goes beyond small changes at home that add up to significant change. People in democracies can also volunteer to be part of the bigger green movement in politics, in industry, or in society.

CASE STUDY: GREEN POLITICS AND ELECTIONS

P olitics is the process whereby leaders are chosen to govern a population by enacting laws, setting policies, and protecting citizens from attack or disaster. A significant part of leadership resides in the ability to project an image that a diverse population understands and accepts. Every industrialized country experiences a period of fast industrial growth in which economic choices often outweigh environmental choices. In the United States, the industrial phase extended into the 1980s. A few decades ago, some leaders used the glib term *tree hugger* to mock environmentalists and win elections. Today few leaders would think of belittling environmental issues. The political pendulum in the United States has swung in the direction of environmental protections, which now make up a crucial part of election campaigns.

Green issues in U.S. politics began in the 1980s, following the lead of the various green parties that had emerged in Europe and other parts of the world. In the 1980s environmentalists known as the greens decided to make an aggressive move into German politics as the Green Party. In 1983, members of the Green Party won seats in Germany's parliament, the Bundestag. Though the Green Party's influence in Europe has strengthened and waned over the years, the rest of the world has taken note that many people accept the protection of the environment as a crucial issue in deciding how to vote.

No government has solved the puzzle of building a strong economy and simultaneously protecting natural resources. Scientists and government leaders often have a difficult time communicating with each other. Few leaders in high government office understand all the intricacies

The following actions are ways for individuals to make a difference in protecting the environment:

- lead by example
- become informed on environmental issues
- vote
- join environmental organizations
- contribute money to environmental organizations
- address public hearings on decisions affecting the environment
- write opinion articles for the newspaper
- run for local office

of ecosystems. Government may furthermore have a hard time accepting the slow process of scientific discovery. Groups such as the Green Party play an important role because their main focus rests in environmental science.

Leaders must find a way to set environmental policies that meet the needs of environmental experts, the public, and industry. All governments should consider the following principles in order to accomplish this daunting task:

- develop an understanding of the environment and ecology
- avoid decisions regarding the environment that cannot be reversed
- act to prevent current environmental damage from continuing
- get tough on polluters regardless of industry pressures and votes from leaders of industry
- integrate decisions so that more than one benefit occurs from an environmental program
- make scientific findings on the environment available to people
- understand that all living things have a right to live on this planet in an environment that does not damage their health.

The activities listed on the previous page are open to almost any citizen of the United States. Some environmentalists have taken more aggressive steps in addition to the ideas listed here. For instance, the following activist approaches to environmentalism are legal and part of the fabric of the United States: volunteering to help eco-conscious candidates for office, organizing company programs for promoting sustainability, writing to elected representatives, joining marches or demonstrations on environmental issues, or training for a career in environmental science.

The green lifestyle does not come with one road map that every person must follow. A green lifestyle is a blend of individual choices that each person makes based on his or her values. The important thing is to do something large or small to help the environment.

CONCLUSION

The current state of Earth's environment contains a single critical factor that outweighs all other environmental issues: In many parts of the world, the human population has exceeded the land's capacity to support it. Industrialized nations in particular live beyond their means by either using up their resources until they disappear or using resources from other places that are not yet overextended. This process cannot go on forever without serious damage to the environment and all of biodiversity.

Green, or sustainable, living originated from the realization that the Earth cannot continue to support people if they reproduce and use up resources at their current rates. Individuals can make green lifestyle choices for reducing wastes, recycling waste, and conserving natural resources. Certain individuals will make a much stronger commitment to preserving the Earth by following the principles of deep ecology. Meanwhile, environmental causes have entered politics and industry after decades in which this topic belonged only to grassroots organizations and a few tree huggers. Environmental matters now make up mainstream politics, lifestyles, and even decisions of large industries.

A large percentage of people in industrialized nations recognize the urgency of improving the environment, yet many of the technologies needed to do this remain in their infancy. Environmental science has made great strides in determining how ecosystems work, but many organisms on Earth have never been identified, so therefore science cannot yet know the entire workings of ecosystems. But people can make a difference by working with the information scientists already know about the environment. First, the world's population has exceeded the Earth's carrying capacity. This makes sustainable activities increasingly urgent. Second, small actions that benefit the environment can have a significant effect when many people take part in them. This is the basic theory behind grassroots organizations that believe change comes from the local level and moves up from there. The environmental movement, perhaps more than any other, can credit grassroots initiatives for its success.

Sustainable activities will certainly depend on new technologies that make it easy for people to preserve the environment. Overall, however, sustainability depends on the leadership from local and national governments plus a commitment from large industries. Most challenging, these groups must learn to communicate with each other and with scientists and environmentalists to achieve success.

Of all sciences, environmental science may depend most on regular citizens to accomplish its goals, either by taking the lead in sustainable living or voting for leaders who do. Sustainability therefore requires more than new technologies. It requires good individual choices, accurate information, and sound leadership.

GREEN BIOTECHNOLOGY

Biotechnology is the use of genetically engineered organisms to carry out specific tasks. Genetically engineered organisms, also called bioengineered organisms, are any microbes, plants, or animals that have received a gene that enables the organism to carry out new activities. The new gene comes from another organism. For example, some species of tomato plants contain a microbial gene that keeps the plants from freezing in cold weather. This is made possible by combining genes that would not normally exist in a living thing. Genetic engineering encompasses all laboratory methods used to carry out this combining of genes.

Green biotechnology consists of genetic engineering for the purpose of solving specific problems in the environment, mainly in the areas of food, new materials, and energy. Despite these benefits, biotechnology has operated under almost constant opposition from people who feel that organisms should not share genes and that any new, bioengineered organism presents unknown dangers to ecosystems.

Jeremy Rifkin has been a critic of biotechnology since the science emerged in the 1980s, and although Rifkin is not completely anti-biotechnology, he nevertheless has spent the past three decades holding the biotechnology industry accountable for safeguarding the public and the environment. Rifkin expressed his concerns to the *Los Angeles Times* in 1991, saying, "This technology is more powerful and more intimate than any technology in history. It's a tool to change the genetic blueprints of living creatures. Scientists have the power to tinker, engineer and rearrange the blueprints of the map of life." In this context, green biotechnology holds a tremendous responsibility.

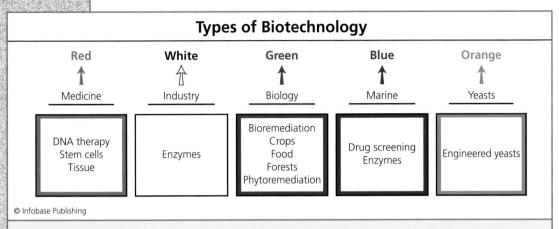

Types of Biotechnology

Red	White	Green	Blue	Orange
↑	⇑	↑	↑	↑
Medicine	Industry	Biology	Marine	Yeasts
DNA therapy Stem cells Tissue	Enzymes	Bioremediation Crops Food Forests Phytoremediation	Drug screening Enzymes	Engineered yeasts

© Infobase Publishing

Biotechnology contributes to the quality of life by leading to new medicines and improving food production. Environmental biotechnology now includes several areas of interest for the purpose of conserving endangered plant and animal species. White biotechnology may deliver the greatest rewards in the future by breaking industry's dependence on nonrenewable and pollution-causing fuels, hazardous chemicals, and energy-demanding operations.

Biotechnology has prompted debate among scientists who create new organisms and a public that worries over the consequences of these new beings in the environment. In addition, technologies sometimes move faster than the speed with which laws appear to regulate them. That has certainly been the case with biotechnology. This chapter explores the promise of green biotechnology in ecology and also explains the current concerns about the safety of bioengineered organisms in the environment.

THE NEW ENVIRONMENTAL SCIENCE

The chemical and food industries, medicine, and agriculture have relied on biotechnology on a large scale since the 1980s, but the evolution of this science unfolded over decades. In 1953 James Watson and Francis Crick described a plausible model for deoxyribonucleic acid (DNA)'s structure. Studies continued on the characteristics of the large DNA molecule for the next several years.

In the 1970s scientists learned to cut DNA into pieces and connect them to create new forms of the molecule. In 1973 biochemists Stanley Cohen and Herbert Boyer laid the foundation for biotechnology by putting one of these new versions of DNA into bacteria. The new organisms became known as *recombinant DNA* organisms. Next, scientists would concentrate on putting specific genes into DNA to carry out desired activities.

In 1980 biochemist Paul Berg was awarded the Nobel Prize in chemistry for introducing techniques in gene splicing, which is the procedure of targeting a specific gene on a DNA molecule and using enzymes to remove that gene from the DNA. In his Nobel Prize address, Berg summarized the state of the art in genetic engineering. (The restriction endonucleases Berg mentioned are enzymes needed for gene splicing. Plasmids are short pieces of DNA that carry a spliced gene into a new organism, and phages are viruses that infect only bacteria and which are also used to transport genes from one type of bacteria into other bacteria.) "Since that time [the 1970s] there has been an explosive growth in the application of recombinant DNA methods for a number of novel purposes and challenging problems. The impressive progress owes much of its impetus to the growing sophistication about the properties and use of restriction endonucleases, the development of easier ways of recombining different DNA molecules and, most importantly, the availability of plasmids and phages that made it possible to propagate and amplify recombinant DNAs in a variety of microbial hosts." All of the components mentioned by Berg now serve as standard tools in biotechnology, described in the following list:

- recombinant DNA—DNA formed with genes from two different organisms
- restriction endonuclease—an enzyme that cleaves the DNA molecule, enabling the insertion of a new gene
- plasmid—a small piece of DNA found in bacteria and useful for recombinant methods
- phage—a virus that attacks bacteria, used in biotechnology to put new genes into bacterial DNA
- amplify—to make many identical copies of a DNA molecule from a single DNA molecule

Entrepreneurs soon began applying recombinant technology to the wide variety of tasks that Paul Berg envisioned. Recombinant organisms have been used to make products for consumers, industry, agriculture, and for uses in environmental science. Recombinant technology has also contributed many advances in medicine, in making new drugs and in dis-

Bioengineering

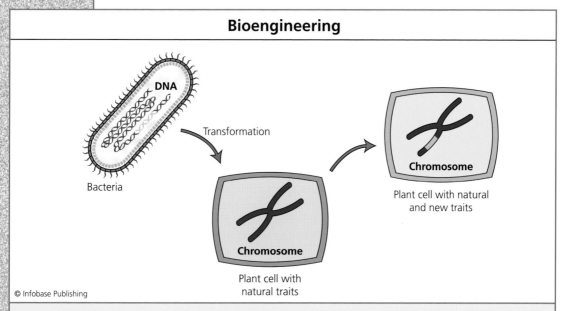

DNA

Transformation

Bacteria

Chromosome

Plant cell with
natural traits

Chromosome

Plant cell with natural
and new traits

© Infobase Publishing

The biotechnology industry owes its success to the process of identifying a desirable gene in one organism and putting the gene into the chromosome of another organism. The new bioengineered organism can then perform an activity it would not normally perform.

ease therapy. Appendix A provides a time line of the events leading up to today's biotechnology.

Green biotechnology is the most recent application of recombinant technology. The following table describes the main uses of recombinant organisms in green biotechnology. Today these products of bioengineering are more likely to be termed *genetically modified organisms* (GMOs), or transgenic organisms.

Bioengineering provides the foundation for all the uses of green biotechnology described here. Each of these tasks uses a plant, an animal, or a microbe containing a foreign gene to enable the organism to perform a new task. Two critical concerns in environmental science today are pollution and climate change, and green biotechnology may be able to help in both of these areas.

Pollution may be obvious or it may be hidden in surface waters, groundwaters, and soils. Bioengineered microbes and plants offer the potential for removing chemicals from polluted places. Microbes or plants can each clean surface soils and surface waters, while bioengineered microbes excel

USES OF GREEN BIOTECHNOLOGY

NAME OF THE TECHNOLOGY	BRIEF DESCRIPTION
agriculture	crops that resist climate changes, drought, invasive species, and require less nitrogen fertilizers
aquaculture	conservation of declining fish species
biocatalysis	microbial enzymes as energy-efficient substitutes for chemically derived products
biodegradation	microbes to degrade pollutants that persist in the environment
bioenergy cells	bacteria or algae for generating energy
biofuels	recombinant organisms to make non–fossil fuels
biopesticides	microbes and plants for nonchemical pest or insect control
bioremediation	microbes for cleaning up pollution
materials science (also biosynthesis)	microbes for new *biodegradable* polymers and plastics
mining	microbes for extracting precious metals from ore
nanotechnology	microbial components for removing pollutants from the body
phytoremediation	plants and trees for cleaning up pollution
recombinant trees	crops that resist climate changes, drought, and disease; efficient producers of wood
waste degradation	microbes for natural breakdown of landfill wastes; wastewater with energy production

at cleaning up polluted groundwaters. New, biodegradable chemicals and pesticides have helped the environment because microbes break them down more readily than they degrade synthetic chemicals, which persist in the environment for years.

Green biotechnology holds the promise of indirectly affecting climate change. New biofuels, new materials, and healthier forests can all result from bioengineering. The European Association for Bioindustries, a leader in green biotechnology initiatives, stated in its 2008 annual report, "Biotechnology is growing from strength to strength. The technology is pervasive and is now being used across many industries. The chemical, pharmaceutical, energy, pulp and paper, agriculture, food industries are all discovering the benefits of biotech which is helping them to make products more ecological, economic and sustainable." Green biotechnology joins renewable energy as two of the most important technologies in building sustainability.

BIOLOGICAL CONTROL OF PESTS

Agriculture depends on plants with traits that result in good crop yields or resistance against bad weather or pests. Pests have plagued farmers throughout history, and chemical pesticides have been used, possibly as early as 4,500 years ago, when farmers sprinkled sulfate powder on crops to ward off insects. Farmers and seamen turned to mercury and arsenic during the Middle Ages to combat insects, mold, and corrosion. By the 18th century, sailors used chlorine for killing pests aboard ships, and farmers experimented with a variety of homemade pesticide mixtures to find a formula that better prevented infestation of crops. In the 1940s and 1950s the United States and Europe began expanding their chemical industries, mainly to support war efforts. This industry served numerous purposes during World War II (1939–45), and at the war's end, chemists turned their attention to new drugs, consumer products, and formulas for agriculture. The period between 1940 and the early 1950s, in fact, became known as the age of pesticides.

Throughout the 1950s chemical companies turned out a steady stream of synthetic organic pesticides containing chlorine, fluorine, or phosphates. These compounds killed primarily insects, which contain eukaryotic cells similar to those of humans. Eukaryotic cells contain membrane-enclosed components called organelles and also hold DNA in

The age of pesticides helped increase crop yields and alleviated hunger in many parts of the world. But the chemicals eventually led to serious pollution and illnesses in wildlife and people. Biological control of insects, pests, and weeds may become the next generation of substances that protect crops but also sustain healthy ecosystems. *(iStockPhoto.com)*

an organized structure called the nucleus (unlike non-eukaryotic bacterial cells, in which DNA floats free inside the cell). Humans, birds, mammals, fungi, and other multicellular organisms contain eukaryotic cells, so the synthetic pesticides that had been invented to kill insects, molds, and weeds also damaged these more complex organisms. In the 1950s the extent of this damage was largely unknown.

In 1962 author and self-trained scientist Rachel Carson published the book *Silent Spring*, which raised the curtain on the potential health hazards of chemical pesticides. Carson's groundbreaking book alerted readers to, among other environmental hazards, the buildup of synthetic pesticides in the environment. She wrote, "For the first time in the history of the world, every human being is now subjected to contact with dangerous chemicals, from the moment of inception until death. In less than two decades of their use, the synthetic pesticides have been so thoroughly distributed throughout the animate and inanimate world that they occur virtually everywhere." Carson had little science to back up these remarks;

she often supplemented her science with intuition, but studies would prove her intuition to be correct.

The public accepted Carson's message quicker than scientists of her day. Science at the time remained focused on new inventions made through chemistry and not on environmental concerns. Carson actually had no proof that pesticides had spread everywhere, yet her words became prophetic by the 1990s, when scientists began detecting traces of pesticides in Antarctica, where no such chemicals had ever been used. In 2008 Mariann Lloyd-Smith, chairwoman of the International Persistent Organic Pollutants (POPs) Elimination Network, teamed with Australian scientists and found large amounts of POPs in the bodies of Antarctica's Adélie penguins: "They [POP pesticides] are incredibly persistent. They stay for many, many, many years, decades, even hundreds of years." During the same time, environmental scientists—environmental science had grown into an important field of study since *Silent Spring*'s publication— were finding an array of synthetic chemicals in addition to pesticides in people as well as wildlife.

In the past 40 years, momentum has swung away from chemical pesticides in most industrialized countries and toward biological means of killing pests. One of the first such biopesticides used in agriculture was the bacterium *Bacillus thuringiensis,* abbreviated Bt. Bt produces a toxic compound that kills insects such as the potato beetle that destroys potato plants. Organic farmers have sprayed Bt directly onto their plants, where the bacteria then produce their insect-killing *toxin.* Bt also produces a tough spore coat around each cell, making it resistant to extremes in temperature, drying, and chemicals. The following sidebar, "Microorganisms from Extreme Environments," describes why bacteria from harsh habitats are an advantage in biotechnology.

The rise of biotechnology as an industry in the 1980s introduced a new way to use the Bt toxin: by putting the toxin gene from bacteria into the DNA of potato plants. The plants then produce their own pesticide to ward off insects. In this case, the Bt toxin is referred to as a *plant-incorporated protectant* (PIP). The Monsanto Company introduced a bioengineered product called NewLeaf potato in 1995, and by 2000 agriculture had accepted it as a successful and safe plant made through recombinant DNA methods. Only one year later, however, the downside of bioengineering emerged. First, the public opposed any type of bioengineering because of fears about its safety, an issue explored later in this chapter in

"Case Study: Concerns about Bioengineering." Second, the combination of the public's concern and the high cost of developing bioengineered plants led NewLeaf to gain no more than 5 percent of the U.S. potato seed market. By 2001 Monsanto stopped further work on the NewLeaf potato.

Garden supply stores continue to sell Bt products to reduce the amount of chemical pesticides applied to home gardens. Since Bt has been introduced as an insecticide, universities have found additional insects that the toxin kills, and Bt remains the only widely used insecticide made by a microbe that is effective against moth and butterfly larvae (caterpillars), potato beetles, mosquitoes in standing water (in flowerpots), fly larvae, black flies, gnats, and corn borers.

Biopesticides offer benefits to the environment in two main ways: (1) They reduce the global amount of chemical pesticides used annually in agriculture and (2) they reduce the home and garden use of chemical pesticides. Home gardeners tend to misuse chemical pesticides, applying too much pesticide over too wide an area. Gardeners can avoid these errors by following the directions on all pesticide containers, including biopesticides. The U.S. Environmental Protection Agency (EPA) controls the use of all pesticides for the purpose of ensuring safety, and this agency requires all products to have clear instructions.

The following lists summarize the advantages and disadvantages of biopesticides derived from bioengineering.

Advantages of Biopesticides:

- less toxic than chemical pesticides
- target only specific pests or types of pests
- do not kill beneficial insects
- nontoxic to people, pets, and wildlife
- usually normal inhabitants of soil
- effective in small quantities
- decompose quickly

Disadvantages of Biopesticides:

- Bt degrades in sunlight and loses activity in 24 hours to a week

- limited benefit on crops attacked by several different types of pests
- Bt must be eaten by an insect in order to kill it
- works slowly; Bt needs a few days before insect populations are affected
- short shelf life; two to three years
- liquid formulations perishable unless kept in cool, dry place out of sunlight
- require knowledge about how biological systems work

Natural biopesticides consist of insects or plants that possess a mechanism to ward off pests. Natural pesticides do not rely on any type of bioengineering and so have been favored by some organic farmers who prefer to use only natural methods for raising fruits and vegetables. Even certain bird species that eat insects can benefit a farm and may be thought of as a type of biological pest-control method. Sustainable agriculture has adopted biopesticides and other natural means of combating insects, molds, and weeds that reduce farm output.

PROTECTING EARTH'S SPECIES

Conservation biology entails new methods of protecting endangered species. Biotechnology contributes to conservation biology because biotechnologists know how to save the genetic content of endangered plants and animals. With this technology, conservation biologists can improve the genetic content of endangered species.

The Zoological Society of San Diego's center for Conservation and Research for Endangered Species (CRES) works on the conservation of wildlife. CRES carries out a program called bioresource banking in which biologists save the genetic material, called *chromosomes*, of critically endangered animals. These animal species exist in such low numbers that they are nearer extinction than other species. CRES scientists keep these chromosomes—usually as semen or ova—in very cold storage conditions called *cryopreservation* (-410°F [-210°C] in liquid nitrogen). Chromosomes stored this way are then available for future studies or for *artificial insemination*.

The principles of biotechnology have helped in the development of *gene barcoding,* or DNA barcoding. Every species has some genes that

Conservation biology focuses on methods of restoring wildlife and plant health. Species exist in almost every place on land and in the ocean, and today's current rapid extinction rates present conservation biologists with an enormous and urgent mission. These scientists may depend increasingly on biotechnology to preserve genes of endangered species and to give species new traits that help them survive in the wild. *(NASA)*

differ from the genes of other species and some that are the same. Within a single species, individuals share many identical genes. Of these identical genes, some are unique like a barcode is unique. The unique genes provide scientists with tools to study populations and subpopulations within a species.

Plant biotechnology takes gene studies a step further by finding genes for favorable traits that will improve plant growth. By putting a foreign gene into seeds or young plants, the plants that mature are said

MICROORGANISMS FROM EXTREME ENVIRONMENTS

Microorganisms that grow in very harsh environments where most other organisms cannot live are called *extremophiles*. Bacteria make up many of the known extremophiles in the world, and the main extremophiles are described in the following table.

EXTREMOPHILE MICROORGANISMS		
TYPE OF EXTREMOPHILE	**GROWTH CONDITIONS**	**EXAMPLE**
thermophile	temperatures to 105–176°F (40–80°C)	*Pyrolobus fumarii*
hyperthermophile	temperatures to 150–230°F (65–110°C)	*Thermus aquaticus*
psychrophile	temperatures to -45–68°F (-7–20°C)	some *Clostridium*
xerophile	very low moisture	*Aspergillus*
barophile	high pressures to 500–1,200 atmospheres	*Colwelia hadaliensis*
halophile	25 percent salt concentration	*Halococcus*
acidophile	pH 3 or lower	*Ferroplasma acidarmanus*
alkaliphile	pH 10 or higher	*Bacillus*
thermoacidophile	temperatures to 176°F (80°C) at pH 1	*Sulfolobus acidocaldarum*
radiation extremophile	radiation up to 1.5 million rads	*Deinococcus radiodurans*
endolith	in rocks and caves	*Acarospora*

(continues)

(continued)

Environmental cleanups can make use of extremophiles because the harsh conditions where some of these microbes live often mimic polluted places. Some microbes actually live in very toxic places. For example, mining technologists search for microbes that flourish in the very acidic mine tailings that flow from mining sites. *Thiobacillus* and *Ferroplasma* live in these conditions, but they produce the acids in mine drainage. By manipulating these microbes' cellular reactions, microbiologists try to invent a new microbe that will clean up the tailings.

Microbiologist Barrie Johnson of the University of Wales said in an interview at a 2005 Society for General Microbiology meeting, "Our ongoing research is focusing on extending the applications of biomining technology, and on using newly discovered extremophile bacteria to simultaneously recover metals and clean up mine effluents from abandoned mines, streams that pass through them and their waste." These natural or bioengineered microbes have been called toxitolerant because they grow in the presence of corrosive acids or bases, organic solvents, toxic metals, or radiation. Said marine biologist Douglas Bartlett of the Scripps Institution of Oceanography in California, "These organisms live in a world that is very different from the skin of the planet in which we humans reside." Science has only scratched the surface of extremophiles' potential in environmental cleanup.

to be bioengineered. Plant bioengineering focuses on the following main focus areas:

- resistance to insects and worms
- tolerance to *broad-spectrum chemical herbicides*
- resistance to infections from bacteria, molds, and viruses
- tolerance to stresses from climate change—heat, cold, drought, increased salt concentration
- delayed ripening
- production of wood alternatives, such as fibers
- production of biodegradable plastics

Aquaculture consists of the confined raising of fish species; it is often called fish farming. Two specialty areas within aquaculture belong to shellfish farming and cultured pearl production. Aquaculture has been used to raise fish that have declined to very low numbers due to either overfishing or changing ocean temperatures, or both.

Aquaculture uses similar principles as used in plant biotechnology, that is, putting a foreign gene into a fish to give the new, bioengineered fish more desirable traits. This new technology is called *blue biotechnology.* The following list presents some of the traits that the aquaculture industry has pursued through blue biotechnology:

- resistance to infections from bacteria, viruses, fungi, and parasites
- increased growth rates
- more efficient conversion of fish diet to food for humans
- tolerance to changing ocean temperatures

Aquaculture's main purpose is to supply regions of the world with a reliable source of seafood or marine products. Aquaculture may also restore depleted natural fishing grounds by giving fish populations a chance to return. Despite the possible benefits, aquaculture draws criticism from two groups of opponents: (1) those who oppose all bioengineering and (2) those who accept bioengineering but fear the effects that fish farming may have on the environment. Bioengineering critics often cite the dangers that bioengineered aquatic organisms cause to ocean ecosystems, much the same way bioengineered plants may cause problems on land. In terms of environmental degradation, fish farming produces large amounts of concentrated wastes, which pollute nearby waters. Dense populations of fish in confinement might also help spread disease, critics contend, and therefore may be less healthy than naturally raised fish.

Fish farming may not be the most efficient way to convert plant energy into an energy source for humans. *Time* magazine writer Ken Stier explored the subject in 2007: "To create 1 kg (2.2 lbs) of high-protein fishmeal, which fed to farmed fish (along with fish oil, which comes from other fish), it takes 4.5 kg (10 lbs) of smaller pelagic, or open-ocean, fish." "Aquaculture's current heavy reliance on wild fish for feed carries substantial ecological risks," says Roz Naylor, a leading scholar on the subject at Stanford University's Center for Environmental Science and Policy. Unless the industry finds alternatives to using pelagic fish to sustain fish farms, says Naylor, "the aquaculture industry could end up depleting an essential food source for many other species in the marine food chain." Aquaculture's detractors and proponents continue to debate the usefulness of this technology.

In 2008 National Oceanic and Atmospheric Administration (NOAA) spokesperson Conrad Lautenbacher assured the public that aquaculture represented an essential part of food production. "The United States has the choice to become an important player in offshore aquaculture to help augment our wild fish products to supply a growing domestic market for healthy seafood." Lautenbacher warned that if the United States did not increase its pursuit of aquaculture, "we will continue to import an increasing amount of foreign aquacultured products, leaving the United States with diminishing control over how our seafood is produced and without the economic benefits from the jobs, technology and innovation that domestic offshore aquaculture would bring." Of course, those same aquaculture jobs would likely grow in regions in the world far poorer than the United States. Aquaculture remains an open question in terms of helping to build sustainable food production.

BIOTECHNOLOGY PRODUCTS

The main products of green biotechnology in use today belong to four categories: agriculture, new materials, biocatalysis, and biofuels. Agriculture represents the largest market in biotechnology products. New crops have been developed by using the plant biotechnology techniques previously discussed. Green biotechnology in agriculture also seeks to make land use more efficient while producing bigger crop yields. These objectives have become increasingly vital as the world's human populations grow at the same time that animal and plant habitats decrease.

Researchers Marshall Martin, Jean Riepe, April Mason, and Peter Dunn of Purdue University's College of Agriculture explained as early as 1996 the need for biotechnology in agriculture: "Over the course of history, scientists, farmers, and food processors have focused on producing food that is more plentiful and higher in quality. As the world's population has grown, however, the total supply of land available for agriculture has remained about the same. In some regions cropland has declined as a result of urban development and/or environmental degradation, such as soil erosion or salinity from irrigation. So the focus of agricultural

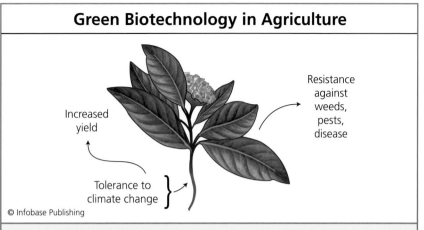

Green Biotechnology in Agriculture

Increased
yield

Resistance
against
weeds,
pests,
disease

Tolerance to
climate change

© Infobase Publishing

To date, environmental biotechnology has had its most important impact on agriculture. About 250 million acres of cultivated land worldwide contain bioengineered crops.

research has been on how to use available land more efficiently and in a more environmentally friendly way." Even with its flaws and ongoing questions of safety, plant biotechnology remains a growing science and an important part of the world's food supply.

A modest amount of research goes into animal bioengineering compared with plant bioengineering. Biotechnology has explored genetic engineering to create leaner meat-producing animals and disease-resistant breeds. At present, no commercial meat products have come from biotechnology.

NANOBIOTECHNOLOGY

Nanotechnology is the science of creating objects that are so small they are measured in nanometers; one nanometer is one-billionth of a meter. The following two examples illustrate the dimensions used in nanotechnology: (1) 6,000 motors designed by nanotechnology could fit on the head of a pin, or (2) a nanometer is like a single marble compared with the size of Earth.

Nanotechnology has been used for centuries, though the term itself is relatively new in science. For instance, centuries-old methods of making stained glass consisted of heating and cooling steps that change the size of glass crystals to nanometer sizes. The smaller crystals then produce different colors in the glass due to the relationship between crystal size and energy. The smaller a crystal becomes, the more the material's electrons are confined to a smaller area. These restricted electrons exhibit higher energy levels and shorter wavelengths than they do in normal material. As a result, the activities of nanomaterials differ from the activities of the same material at gram or even microgram amounts.

In 2007 University of California nanotechnology scientist Jeffrey Grossman explained to KQED television in San Francisco, "The behavior of nanomaterials changes, or can change, when the size becomes so small when compared with a larger amount of that same material." Nanotechnology seeks to customize these unique behaviors to perform tasks that were previously too difficult or too expensive to perform by conventional science. Three primary areas under study today in nanotechnology are: (1) nano-sized computer chips that run faster than current chips; (2) tiny medical devices that work inside the body's vessels; and (3) new filters with nanoscale pores for filtering contaminants out of drinking water.

Nanobiotechnology uses techniques at the nanoscale to study biology or to develop new biological processes. The main focus areas in nanobiotechnology today are listed in the following table, along with possible ways in which this technology can be used in environmental science.

Nanoscale objects present some safety concerns that science has not fully defined. Nanoscale particles are so small that they have the potential of infiltrating the body, especially by inhaling. Any new science has experienced in its history inevitable phases of doubt, safety concerns, opposition from different scientific viewpoints, and acceptance or rejection by the public. Nanotechnology will also travel this road. The current concerns in nanotechnology are the following:

- new types of pollutants from nanomaterial manufacturing
- biological harm to organisms by highly reactive nanoparticles
- possible increase in toxicity of some chemicals when reacted with a nanomaterial
- nanomaterial disruption of activities in ecosystems
- nanomaterials entering food chains
- possible accumulation of nanomaterials in human and animal tissue or in plants

Nanobiotechnology is at its leading edge as a science, so it holds far more questions than it has answers. Obviously, the first item to address is

NANOBIOTECHNOLOGY IN ENVIRONMENTAL SCIENCE		
RESEARCH AREA	**DESCRIPTION**	**POSSIBLE USES IN ENVIRONMENTAL SCIENCE**
analytical systems	new fluids produced by nanotechnology used in combination with current analytical methods, such as *spectroscopy*	detection of minute amounts of pollutants in the environment and in organisms
biocatalysis	chemical reactions catalyzed by new, safer materials	avoidance of toxic catalysts needed by current chemical processes; neutralization of pollutants
biosensors	probes used for detecting substances in biological systems, usually at very small quantities or sizes	detection of reactions in the environment or in organisms that may lead to toxic conditions
cellular communication and cell surface activities	nanoscale analyses of the junctions between living cells and the structures on the outside of cells	learning how chemical or biological pollutants disrupt cell and organ function, leading to disease
health	studies of the health impacts of nanomaterials	assuring that new nanomaterials are not dangerous to human and animal health
manufacturing	processes using nanoscale reactions at ambient temperature and pressure	manufacture of products and materials using less energy demand than conventional manufacturing

(continues)

NANOBIOTECHNOLOGY IN ENVIRONMENTAL SCIENCE *(continued)*		
RESEARCH AREA	DESCRIPTION	POSSIBLE USES IN ENVIRONMENTAL SCIENCE
nanofabrication	studies nanomaterials and their behaviors at surfaces and in unique reactions	new energy-producing devices, new materials that do not deplete natural resources
nanotubes	nanoscale tube structures that can conduct electricity	avoidance of toxic materials in current electronic devices and semiconductors
tracking materials	nanoscale materials built into products and industrial chemicals	for tracking industrial products from manufacture to disposal

whether nanomaterials present more danger to the environment than the potential benefits they deliver. But the potential benefits of nanotechnology and nanobiotechnology are so enormous that science seems obligated to pursue them, especially in the search for new types of energy. Richard Smalley, winner of the 1996 Nobel Prize in chemistry, told *National Geographic* magazine in 2006, "This is the great getting-up morning of nano. If Mother Nature allows it, we could restring the electrical grid of the world." Smalley's optimism may signal a new generation of nano-produced energy, perhaps based on the restructuring of molecules such as carbon to carry currents. The nanotechnologies may have a very promising future.

BIOENGINEERING'S IMPACT ON BIODIVERSITY

Bioengineering has the promise of preserving biodiversity if it helps clean up pollution, restore habitat, degrade wastes, contribute to reversal of global warming, and develop materials that conserve natural resources. In fact, all of these topics are areas of study in environmental biotechnology.

Depletion of natural resources leads to biodiversity loss by ruining habitats needed by plant and animal species that are under threat of extinction. Several aspects of green biotechnology foster better care of biodiversity by preserving habitat and the natural resources now consumed by humans. Biodegradation, bioremediation, and *phytoremediation* all represent bioengineering-based sciences that are intended for pollution cleanup. Bioenergy cells may soon contribute as well by offering alternative energy sources to replace batteries containing toxic metals, and bioengineered trees may help conserve natural forests.

Bioengineered crops carry advantages and disadvantages. Properly used, bioengineered crops need less chemical fertilizer and water, grow quickly, resist pests, and provide nutrition in areas of the world with poor growing conditions. Critics fear that these features are accompanied by new allergies, new toxic substances, and the unpredictable effects of unnatural genes in the environment. *(iStockPhoto.com)*

CASE STUDY: CONCERNS ABOUT BIOENGINEERING

Bioengineering bypasses many of the normal processes in which parents transfer their genetic history to their offspring. Bioengineered organisms contain genes for traits that might never occur by *natural selection*, and these traits certainly would not develop with the speed in which bioengineering creates them. This sense of toying with nature has made many people and even many scientists wary of bioengineering. The conservation group Sierra Club has objected to bioengineering, as have other environmental organizations. In 2001 the Sierra Club stated, "Genetic engineering is not, as many of its supporters claim, merely a more efficient form of plant and animal breeding. There is a clear boundary between traditional breeding methods and the radically new technology of genetic engineering." Genetic engineering has, in fact, spawned one of science's most heated debates.

Opponents propose a number of reasonable fears regarding biotechnology. No process in science is perfect, and bioengineered organisms may behave in ways that were not intended. Critics' main arguments against bioengineering are the following: (1) The unnatural organisms may cause unanticipated harm to natural organisms; (2) the effects of a new gene may wear off over time (insects resistant to plants engineered to repel those insects may develop, for example); (3) the genes may transfer to other organisms with unknown effects on the environment; and (4) people may develop of new allergies to bioengineered plants. The Sierra Club added, "The accidental flow of . . . altered genes from a genetically engineered organism to a natural organism, by pollen transfer or by other means, results in the production of an organism which, although it has not intentionally been genetically engineered, must be classified as a genetically engineered organism." These issues have yet to be resolved.

Environmental groups have lined up with the Sierra Club with their own arguments against bioengineering, while proponents try to argue the technology's benefits. Greenpeace has called

Climate change affects the sustainability of forests, grasslands, polar regions, oceans, and every other habitat or ecosystem that supports biodiversity. The ultimate gift that green biotechnology can give to the world would be a way to directly or indirectly affect climate change. Plant biotechnology in agriculture may turn out to be the biggest contribution from this science because agriculture must expand to feed a growing world population. Since the most fertile agricultural lands are already being used, cropland now spreads into grasslands and forests, the two most threatened *biomes* on the planet. Green biotechnology can arrest this trend with bioengineered crops offering higher-than-normal yields per acre of land.

into question whether bioengineered crop yields exceed regular crop yields and whether the bioengineered varieties reduce chemical herbicide use. Greenpeace also fears that herbicide-resistant "superweeds" can emerge from cropland growing plants engineered to resist weeds. New Zealand's Greenpeace warned in 2004, "Consumers the world over are well aware that GE [genetically engineered] crops offer them no benefits and continue to demand GE-free foods. GE has been given a huge 'chance' in the United States and has proved itself an environmental, public, commercial and agronomic failure." The agricultural organization CropLife International countered in 2006 by explaining the value of bioengineered crops: ". . . small-scale farmers [in developing countries] tend to benefit most from biotech crops, as insect and disease-protected crops provide new and previously unavailable tools to combat pest problems. These crops are delivering major benefits to farmers and society, through increased yields, higher incomes, simplified crop management, and, in some cases, reductions in the use of pesticides." Any member of the public could not be blamed for becoming confused about whether biotechnology brings great benefits or great dangers to the environment.

Bioengineering has made astounding progress during its short history. New drugs made by biotechnology companies improve human health. But bioengineered drugs do not enter the environment the same way that bioengineered organisms potentially can, so the public is less worried about medical biotechnology. Medical and environmental biotechnology are still new sciences compared with chemistry or biology. Any new science holds the potential of bringing harm to society unless it is treated with respect. But with trained biotechnologists, sound experiments, and safety steps developed specifically for engineered organisms, bioengineering will likely play a role in sustainability.

Forests help reduce global warming by storing a major greenhouse gas, carbon dioxide (CO_2). In this way, forests help reduce the overall buildup of greenhouse gases and the rise of global temperatures. Today's forests have been depleted in some parts of the world, however, so that they cannot remove atmospheric CO_2 in quantities equal to the amounts produced by vehicles and deforestation. Conserving the present forests and rebuilding much of the depleted forested land would help in removing CO_2 from the atmosphere and thus help control global warming. New bioengineered trees represent just one example of how green biotechnology can help conserve natural forests and therefore play a part in conserving biodiversity.

CONCLUSION

Green biotechnology is a specialty in the larger field of biotechnology that focuses on ways to improve the health of the environment. This science specifically uses the transfer of genes among different types of organisms to achieve its goals. Agriculture currently represents green biotechnology's main emphasis. In this arena, scientists create food-producing plants that produce higher-than-normal yields and plants that resist damage from pests.

Technologies that create bioengineered organisms offer opportunities to improve ecosystem health, reduce pollution, and protect biodiversity, but scientists cannot make quick progress when they must simultaneously defend their science. Biotechnology has already made inroads in conserving natural resources, decreasing chemical pesticide use, improving food production in areas stricken by the effects of global warming, and devising new ways to produce energy. Nanobiotechnology may soon follow with advanced energy-making processes and microscopic devices that combat environmental diseases. These breakthroughs depend on equal amounts of science and support from the public.

Of almost all the new technologies created in the past century, bioengineering has created a tremendous amount of concern along with its promise. Many of the arguments against new technologies may be well founded, but scientists and nonscientists must work together to differentiate real from perceived threats. The success of green technology therefore rests with two complementary actions. First, scientists must make every effort to explain their technology, its benefits, its safety, and also any safety concerns. Second, the public must hear science's message and make informed choices about the technologies they support and do not support. Perhaps critics of biotechnology, for instance, will serve key roles in assuring that this growing science creates inventions that are both effective and safe for the environment. In the end, everyone must realize that we are all on the same team trying to conserve natural resources for future generations.

SUSTAINABLE AGRICULTURE AND BIOPESTICIDES

S ustainable agriculture has become synonymous with organic agriculture or organic farming because both seek to protect the environment while producing healthy food. Today organic farming, which is the growth of crops or food animals without using pesticides or artificial growth enhancements, usually consists of small-scale farming that serves local markets. Sustainable agriculture uses the same principles but on a much larger scale. Sustainable agriculture strives to produce the same healthy foods that are produced by small farms but also to change the way farming affects the global environment. Sustainable agriculture seeks to lower the use of fossil fuels, conserve water and energy, and lessen or eliminate the use of pesticides and chemical fertilizers—all on a national or global scale. In the United States, the Department of Agriculture (USDA) supports the decision to convert traditional farms to sustainable farms, and it offers courses on sustainable practices.

This chapter describes the current practices in sustainable agriculture and one of its most important tools, biopesticides. The chapter describes the sustainable agriculture movement and investigates the workings of a typical sustainable farm. It also examines special areas in sustainable agriculture in addition to biopesticides, such as specific bioengineered crops and methods for keeping bioengineered materials from harming the environment. Chapter 3 closes with a discussion on the international efforts in sustainable agriculture and highlights how some countries have advanced past the United States in building sustainability.

THE SUSTAINABLE AGRICULTURE MOVEMENT

At the end of World War II in 1945, the United States and other combatants switched their manufacturing focus from military equipment to peacetime machinery. Agriculture inherited the task of feeding a rapidly growing postwar population called the baby boomers. Farming had already been evolving from family-owned parcels to huge commercial operations, but the baby boom accelerated this change. Agriculture adopted mass production, large mechanized equipment, and large-scale chemical use.

Sustainable agriculture seeks to reverse the trends of large corporate agriculture by using only methods that cause the least damage to the environment or possibly improve the environment. Sustainable agriculture has two additional objectives: (1) to earn a profit and (2) to enhance the lives of people who choose sustainable methods. Sustainable agriculture, if allowed to develop worldwide, may also help reduce malnutrition. Sustainable agriculture is less harmful to soil and conserves water. This enables impoverished areas to grow crops for several seasons without depleting the land and allows people to grow more food for a longer period of time than would be possible with more wasteful types of agriculture.

Sustainable agriculture is also referred to as low-input agriculture because it consumes lower amounts of energy, water, and nonrenewable resources than conventional agriculture. The three main focus areas to achieve low-input goals are water conservation, soil conservation, and pest control. Water conservation involves the use of more water-efficient crops and irrigation systems while decreasing reliance on groundwater and surface water and reducing water waste. Sustainable agriculture also promotes a decrease in meat production, which uses water inefficiently, and a shift in diet toward more energy- and water-efficient foods. Soil conservation involves the increased use of perennial crops, crop rotation, and organic fertilizers and landscaping, ground cover, and the limited use of tilling. To conserve soil, sustainable farming also tries to reduce overgrazing, road-building, and irrigation that depletes natural waters. Polyculture, the growth of mixed crops, also serves to conserve water and soil.

Pest control represents a key part of any farming operation. Conventional farming hurts ecosystems by putting large amounts of chemical fertilizers into the soil and tons of organic wastes into surface waters

Sustainable agriculture has gained proponents in the United States because it supports the environment while decreasing costs for growers. A well-managed sustainable farm reduces the following items and therefore cost: fuel, energy, chemical fertilizers, pesticides and herbicides, water, and soil erosion. The field shown here has been tilled, but no-tillage farming may be better for keeping nutrients in the soil for next season's crop.

and often uses practices that cause soil erosion. Most important, conventional farming has long depended on chemical pesticides. The chemicals from pesticides, fertilizers, and animal waste each in their own ways can destroy ecosystems.

Agricultural chemicals poison many of the living things in ecosystems in two main ways. First, organic wastes pollute surface waters with high levels of nitrogen and phosphorus, which lead to a rapid overgrowth of water microbes that deplete the water's oxygen. The process in which dense microbial growth removes all the oxygen from the water is called *eutrophication,* and this situation is deadly for fish and other organisms that live in the water. Second, chemical pesticides interfere with animal reproduction, growth, and overall health.

New farming methods minimize or eliminate the use of chemical fertilizers and chemical pest control. Sustainable farms now experiment with a combination of methods for growing healthy crops while still inhibiting weeds, insects, and disease-causing microbes. This combination of growing methods and pest control methods is called *integrated pest management*.

A SUSTAINABLE FARM

The best sustainable farms incorporate integrated pest management, water and soil conservation, and energy conservation. Sustainable agriculture has different ways of achieving pest control and natural resource conservation, so if visiting all the sustainable farms in the world, a person would find that few farms do things exactly alike.

Integrated pest management on such a model farm would use one or more of the options described in the following table. This type of pest control is called integrated management because the pest control program relies on complementary or integrated methods, such as natural predators and natural insecticides combined with planned crop rotations.

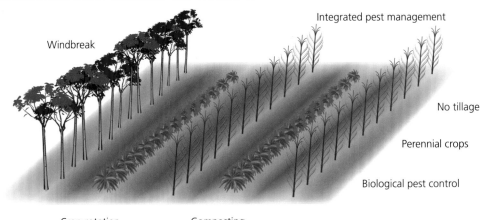

Sustainable Farm on Flat Terrain

Integrated pest management

Windbreak

No tillage

Perennial crops

Biological pest control

© Infobase Publishing   Crop rotation   Composting

Sustainable farms work with the local environment rather than try to tame nature. On flatland, fields should be oriented toward the Sun and protected by trees that reduce wind erosion. No-tillage farming, the use of composted materials, and seasonal crop rotation help replace nutrients in the soil, thus reducing the need for fertilizers. Perennial crops that regrow for more than one season help conserve fuel and energy required for seeding fields.

INTEGRATED PEST MANAGEMENT		
PEST CONTROL METHOD	**DESCRIPTION**	**EXAMPLES**
bioengineered crops	insecticide gene put into the DNA of crops	insect-resistant potato plants
biological control	insects in croplands that combat pest insects	ladybugs eating aphids
biopesticides	biological controls, usually microbes, to kill pests on plants	*Bacillus thuringiensis* bacteria
insect sterilization	chemicals or radiation to make pest insects unable to breed	insect hormones
natural insecticides	insect-killing substances produced in nature	pyrethrin compound made by chrysanthemum plants
natural predators	insect-eating birds or other animals that prey on pest insects	flycatchers (birds) and bats
polyculture and crop rotation	variation in crops to reduce populations of crop-specific pests	alternating rows of apple trees, peach trees, and raspberry bushes

Sustainable farming puts substantial efforts into conserving natural resources in the same way it emphasizes the use of nonchemical pest control. Natural resource conservation targets the most efficient water-use and soil-use methods available in farming. The following table describes these approaches, listed in general from the least efficient at the top to the most efficient.

The most effective water and soil conservation programs use a blend of the methods discussed here. For instance, drip irrigation in combination

NATURAL RESOURCE CONSERVATION IN SUSTAINABLE AGRICULTURE	
CONSERVATION METHOD	DESCRIPTION
Water Conservation	
furrow irrigation	small channels or furrows dug between rows of crops for delivering water in flat terrain
sprinkler systems	overhead sprinklers spray water directly onto crops in hilly or flat terrain
subirrigation	underground tubes deliver water to the roots
drip irrigation (trickle irrigation)	tubes deliver water directly to the root area
Soil Conservation	
shelter belts	row of trees planted on cropland perimeter to decrease wind exposure
strip-cropping	plowing land perpendicular to the greatest wind exposure
elimination of erosion-causing irrigation methods	avoidance of flood-style irrigation or delivery systems that cause runoff on hilly terrain
contour farming	soil plowed according to the land's natural slope and shape
terracing	landscaping hilly terrain to slow runoff on hillsides
no-tillage, low-tillage farming	reduction of plowing and tilling to prevent loosening of soil and wind erosion
erosion-resistant ground cover	planting clover or grasses between or near crops to stabilize the soil

Sustainable Farm on Hilly Terrain

Contour or terraced planting

Efficient irrigation

Organic fertilizers

Biological pest control

Soil conservation

Water conservation

© Infobase Publishing

Sustainable farming on hilly terrain must minimize soil and water loss. Terraced planting uses flat strips of land built on a hillside; contoured planting uses less uniform flat areas built to follow the hillside's natural contour. Terraced and contoured fields contain ditches that capture rain and irrigation runoff. Drip irrigation works better than spraying. Like farms on flatland, hilly farms use biopesticides, natural fertilizers, perennial crops, and minimal tillage.

with contour farming might be the best choice for hilly land with little wind exposure. Flat agricultural land in dry, windy regions might do better with a combination of furrow irrigation, shelter belts, and planting ground cover plants.

No-till farming, in which farmers do not turn over the soil in their fields between growing seasons, delivers several benefits to the environment and to sustainable farming, as follows:

- decreases wind and runoff soil erosion
- decreases equipment fuel use and emissions
- with crop rotation, reduces use of fertilizers
- improves wildlife habitat

In no-till farming, farmers leave plant residues in the soil following a harvest and do not plow or disk the stubble back into the soil—a practice known as reduced tractor trips. The residues break down naturally and enrich the soil with nutrients. No-till farming often employs weed-control methods so that weeds do not overgrow the fields. This weed control usually consists of planting ground cover to inhibit weed growth. In areas where weeds cause a particular problem, farmers might use a small application of chemical herbicide on bioengineered, herbicide-tolerant crops. Bioengineered crops and simultaneous no-tillage farming also provide an economic benefit, explained by Almir Rebelo of the Brazilian grower organization Friends of the Earth: "Our problem with erosion was very serious and it was very damaging to the environment to the extent that . . . to produce 1 [metric] ton of grain in Brazil, we lost 10 tons of soil per hectare per year. We solved this problem by eliminating tillage." Even conventional farms are beginning to experiment with no-tillage farming.

Sustainable farming requires good planning so that the land produces a profitable harvest yet the environment receives minimal damage. The U.S. agricultural industry has been slow to accept sustainable methods that other countries have already tried, perhaps because of misconceptions about sustainable methods. Some of the questions surrounding sustainable farming come from the following concerns or misconceptions:

- Crop rotations or fallow fields (seasonal uncultivated fields) are not economically feasible.
- Sustainable methods cannot keep up with global food demand.
- Sustainable and organic farming are feasible only for the "alternative foods" market.
- Government regulations will soon enter sustainable farming so that farmers will lose their freedom of choosing methods that work best for them.

The sustainable movement in agriculture has already addressed the concerns mentioned above, as is highlighted in "Case Study: Australia Balances Its Natural Capital," which follows. The environment may have reached a point in which tough choices must be made in order to feed a growing population but also preserve as much habitat as possible. Sustainable agriculture will undoubtedly play an important role in meeting this goal.

Wildlife living near no-tillage operations has a chance to rebound, probably because heavy equipment no longer traverses the fields two to three times before each planting. Bioengineering proponent and grower

Ecological styles of agriculture operate in ways to make as little impact on the surrounding ecosystems as possible. This may include planting fields interspersed with natural meadows, leaving trees standing, and encouraging native predator and prey species to live in and around the cultivated fields. *(USDA)*

Jay Hardwick mentioned in an interview with Monsanto in 2006, "As a result of us keeping crop residue on the ground, we have a new foraging opportunity for wildlife. So we're seeing a new happening on the landscape in terms of wildlife emergence. Not only on top of it, but underneath. Earthworms are coming back to play, and earthworms are strategic in getting water into the soil structure." No-tillage therefore enhances the environment along with its other known benefits.

Sustainable agriculture's success may someday be measured by its progress in contributing to two new concepts called *agroecology* and *agroecosystems.* Agroecology is the science of applying ecological concepts to the design and management of sustainable farms. In other words, sustainable farms become part of agroecology when the farmers do all they can to reduce the farming operation's ecological footprint. An agroecosystem is an agricultural system that also acts as an ecosystem. The conditions described above by Jay Hardwick represent an agroecosystem.

(continues on page 64)

CASE STUDY: AUSTRALIA BALANCES ITS NATURAL CAPITAL

In 2007 the Reuters news agency released an update on the current and future status of the world's agriculture. The news release stated, "Overall, the world's agricultural productivity was forecast to decline between 3 percent and 16 percent by 2080, according to the study published by the Washington-based Center for Global Development and the Peterson Institute for International Economics. Among developed countries, Australia's outlook was bleakest with predicted declines in crop yields ranging between 16 percent and 27 percent." Australia has become a test system for how best to manage inhospitable environment with agriculture.

Australia is an island nation containing mostly dry land that supports little cultivation compared with other lands its size. The country is vulnerable to drought, water wastage, and climate change. To address the strong connection between Australia's limited resources and its agricultural needs, the country has taken a leading role in exploring sustainable and organic methods of raising crops. For instance, an organic certification group named the National Association for Sustainable Agriculture, Australia (NASAA) has been operating for more than two decades to meet Australia's agricultural challenges.

In the recent past, Australia has seen a loss of grasslands, grazing land, and water supply while experiencing decreasing water quality, increased salinity (salt content) of the water, and drought. The agricultural arms of Australia, grazing animals and cropland, have turned to sustainable methods for halting further destruction to the land. Farmers in dry Western Australia can probably see the end of their livelihoods on the horizon if they do not switch to ways of working with the land rather than trying to tame Mother Nature. Grain and sheep producer David Curtin was quoted in a Department of Agriculture and Food brochure, "Farming for the Future": "We are closely monitoring water through harvesting; stopping erosion and waterlogging. We have also been planting trees. We have three natural creeklines that have been revegetated. This has been good for production by reducing wind erosion and preventing salinity." Curtin could have also added that these techniques are fairly easy and inexpensive.

Australia offers an example of a nation that may be driven to sustainable methods out of necessity to save a vulnerable environment. Parts of Australia experience violent swings in weather, from drought to heavy rainfalls; water, soil nutrients, and carbon are at a premium in the country's undeveloped outback. Australia's rivers consist of mostly slow-moving water that washes only small amounts of salts and sediments away. This slow flow also returns few nutrients to the soil, making farming even more difficult.

Grazing and cultivation that worked well enough in years past in Australia's dry midcontinent will no longer work in the near future. The only way rural grazing or farming has traditionally scratched out a living has been by making radical changes to the *water cycle* and the *nutrient cycles* of the land. The country is therefore embarking on new systems for creating sustainable farms tailored to its unique environment and climate. Agriculture landscape experts John Williams and Hester Gascoigne wrote a 2003 article, "Redesign of Plant Production Systems for Australian Landscapes," for the Australian Society of Agronomy that described the new steps that should be undertaken on Australia's continent. Their main suggestions were the following:

- selection of soil, plants, and animals that act together as an ecosystem
- building of paddocks, barns, and other buildings with least disruption to existing ecosystems or which do not hinder reestablishment of an ecosystem
- soil, nutrient, and water conservation methods best suited for the region
- replanting native vegetation to stabilize soils and underground water sources
- recharging underground water sources with freshwater to decrease salinity
- planting trees over or near high-salt areas to keep salts from migrating into crop or grazing land
- crop selection to reduce nutrient loss from soils or planting of compatible species that, overall, conserve nutrients

Australian farms have also been among the first in global agriculture to experiment with *companion farming* in which annual cereal seeds are planted on perennial grasslands during the grass's dormant months. This reduces water loss and soil erosion between seasons. The most promising companion plants tend to be plants with root systems that hold nutrients and water in place and reduce their loss.

The following five technologies may become good choices for Australia's dry sustainable farming and companion farming: (1) repressing the flowering gene in cereals to keep it in a constant growing and root-developing stage; (2) bioengineering grain legumes to be perennial growers so they store soil nitrogen year-round; (3) bioengineering plants for higher transpiration rates, which helps nutrient cycling; (4) breeding new plant types that are adapted to Australian growing conditions; and (5) integrated pest management to improve the efficiency of crop yields.

(continues)

(continued)

At the 2008 Farmer of the Year award presentation in New South Wales, Australia, Minister for Primary Industries Ian Macdonald pointed out, "The fact that all of this year's finalists, as well as most of the overall applicants, are battling the drought and other financial pressures but are still tackling environmental challenges is indicative of the determination and talent across our state's farming sector." Australia's farmers are showing the world that sustainable farming is not a luxury but a necessary way to sustain their way of life.

(continued from page 61)

BIOPESTICIDES

Biopesticides come in the three following forms: (1) microbial biopesticides produced by bacteria, fungi (usually molds), or viruses; (2) plant-incorporated-protectants (PIPs), which are substances made by plants themselves to ward off pests; and (3) biochemical pesticides that consist of substances made in nature, which have activity against pests. In this context, "pests" refers to insects, larvae, worms, and disease-causing microbes.

The most common biopesticide is a toxic protein made by *Bacillus thuringiensis* bacteria, or Bt. This crystal-shaped protein binds to the lining of the insect larva's gut after being ingested and causes the larva to starve to death. Additional biopesticides have been studied in biotechnology in which a desirable gene from one microbe is put into another microbe that grows well on crops. The new microbe protects the crop from insects or disease.

Agricultural bioengineering offers the advantage of creating a microbe to do a specific job, but it must overcome the same public opposition that occurs in other biotechnologies. For this reason, agriculture biotechnologists have developed an effective safety mechanism for bioengineered crops, described in the following sidebar "Apoptosis—Suicide Genes for Bioengineered Bacteria."

Many useful and natural microbes kill insects or carry out other jobs without the need for bioengineering. For example, *Pseudomonas fluore-*

scens bacteria help reduce frost damage on certain crops. The table below describes some of the promising natural biopesticides and examples of the insects they kill.

Users of biopesticides must remember that biological products do not act as quickly as human-made chemicals. Also, biopesticides have narrow ranges of activity, meaning they usually act on a small number of insect species. (Bt is an exception that attacks a wide array of insects.) Biopesticides are safe for humans and animals, and they break down in the environment after a few weeks because they are biodegradable substances.

The newest generation of biopesticides uses bioengineering to make PIPs in which scientists take a gene from an organism and put it into an unrelated plant, giving the plant the ability to repel insects. The gene for the Bt protein has been one of the most popular sources of insecticide

EXAMPLES OF BIOPESTICIDES		
MICROBE	**TYPE**	**BIOPESTICIDE TARGETS**
Agrobacterium	bacteria	other types of disease-causing *Agrobacterium*
Bacillus pumilus	bacteria	molds
Bacillus thuringiensis	bacteria	beetle larvae, budworms, caterpillars, corn borers, mosquitoes
Beauveria bassiana	fungus	caterpillars, Colorado potato beetle, mealybugs, weevils, whiteflies
Metarhizium anisopliae	fungus	root weevils, sugarcane froghopper
Verticillium lecanii	fungus	aphids, jessids, whiteflies
Muscador	mold	other molds
baculoviruses	virus	butterflies, moths
Iridoviridae	virus	beetles, flies

activity for these PIPs. The gene then stays with the plant species through several generations so that new young plants also repel the pest.

Biochemical pesticides act in three main ways without the need for the bioengineering needed to make biopesticides. (Biochemicals are chemicals made by natural processes.) First, some natural substances kill pests. For example, many herbs, such as oregano or rosemary, produce oils that inhibit disease-causing microbes. Second, natural substances sometimes interfere with insect hormones and so disrupt insect mating. Insect *pheromones,* for example, lure insects to seek a mate that does not exist and disrupt normal reproduction. This, of course, can reduce the numbers of insects in an area in a single growing season. Third, some biochemicals attract insects to traps. Pheromones can be used in this way also to draw insects into mechanical traps. Wasp traps use a sugar-decayed protein mixture to attract the wasps into a simple container that acts as a lethal trap.

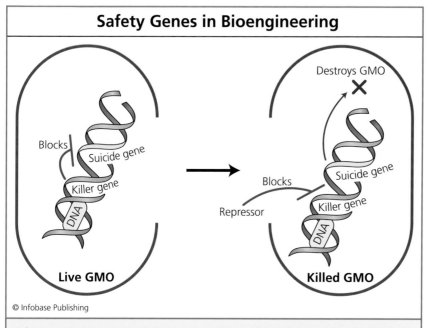

Safety Genes in Bioengineering

© Infobase Publishing

Safety genes in bioengineered plants or microbes speed the organism's natural demise called apoptosis. In the environment, a genetically modified organism (GMO) may carry out its bioengineered activities, but after a certain period, the safety genes activate and shut down further propagation of the GMO. This helps keep unnatural genes from entering the environment in an uncontrolled manner.

APOSTOSIS—SUICIDE GENES FOR BIOENGINEERED BACTERIA

Apoptosis is the natural and organized death of any type of cell: plant, animal, or microbial. In multicellular organisms—mammals, for example—apoptosis occurs continually as old cells shut down their processes and new cells take their place. A variety of known and unknown compounds trigger apoptosis in microbes. Scientists have used this information in bioengineering for the purpose of making genetically modified organisms (GMOs) that do not persist in the environment. By manipulating apoptosis in GMOs, biotechnology can assure the public that the GMOs will not cause havoc in the environment.

Genetic engineers have learned how to put a so-called *suicide gene* into a bioengineered plant or microbe. Suicide genes initiate apoptosis in cells, and they have been shown to work in food plants. The plants grow only for one growing season and do not germinate the next season. (In order to use bioengineered plants containing traits that farmers want, a farm must buy a fresh supply of bioengineered seeds with the suicide gene each season. Unfortunately, this idea would be too expensive for the majority of farms in developing countries and possibly an unreasonable expense for farms in rich countries.)

Suicide genes can also enable bioengineered bacteria to work safely in the environment. These bacteria contain a clever bioengineering scheme that includes both a suicide gene and a killer gene that deactivates the suicide gene. Biotechnologists first devise a killer gene that remains inactive in the presence of a particular pollutant. As long as the pollutant exists, the bioengineered bacteria degrade it. But as the pollutant breaks down into harmless compounds, the killer gene turns on. The killer gene activates the suicide gene, which then destroys the bacteria. This scheme allows bacteria to do their job, then disappear.

The entire switch-on and switch-off process consists of three components: (1) the suicide gene, (2) the killer gene, and (3) a gene called a protector gene, which turns the killer gene on or off. Scientists can create protector genes that respond to conditions in the microbe's immediate environment. In this way, scientists can devise microbes for specific

(continues)

(continued)

pollutants. In 1988, when suicide and protector genes were first being developed, microbiologist Stephen M. Cuskey of the Environmental Research Laboratory in Gulf Breeze, Florida, said to the *New York Times*, "Theoretically, there are lots of ways to control the protector gene. It can be designed to respond to different temperatures, or light or darkness, or to other chemicals. It's a versatile system." Control of apoptosis allows farmers to use GMOs with far more safety assurances than in the past.

GMOs that contain apoptosis controls may help in solving the chemical pesticide problem than has grown worldwide since the 1960s. Chemical pesticides create two potential hazards in ecosystems. First, chemical pesticides eliminate natural pest control by harming the health of predator species, and second, they may lead to the creation of pesticide-resistant insects and parasites. GMOs with apoptosis controls offer an efficient and safer approach to pest management.

PLANT-INCORPORATED PROTECTANTS

Bioengineering can now design plants with specific pesticides in their leaves or other plant parts. The plants themselves produce a pest-repellant substance after a new gene has been incorporated into their DNA. The altered plant then produces more pesticide as its cells reproduce and grow. Often, seed suppliers simply put the new gene into a plant's seeds so that all the new seedlings contain the pesticide.

Three main technologies work best for producing PIPs: (1) Bt bacteria, (2) CryIA(c) toxin gene, and (3) mosaic viruses. Bt bacteria deliver well-known benefits as whole cells sprayed onto crops. Bt also serves as the source of an insecticide gene, which is put into plants by bioengineering. This Bt gene, known as CryIA(c), controls the manufacture of Bt's natural pesticide, called delta-endotoxin. Genetic engineers have put CryIA(c) into other bacteria and into plants, and it is now used more in green biotechnology than whole Bt bacterial cells.

Mosaic viruses offer a more efficient way than CryIA(c) to get new genes into plants. This is because select viruses naturally infect plant cells by injecting their genes directly into the cell where they mix with the plant

DNA's genes. By putting a desirable gene into a virus, viruses save scientists the time and trouble of forcing foreign genes into plants.

Five main types of infective mosaic viruses have been investigated for making PIPs: (1) watermelon mosaic virus, (2) squash, (3) zucchini, (4) papaya, and (5) cucumber. Scientists must first deactivate the lethal parts of these viruses before putting new genes into them.

PIPs can also come from new technologies in plant vaccines. *Newsweek* writer Geoffrey Cowley described the method in 2001: "Researchers have made inroads against crop-killing viruses and bacteria, but they're using a different technique. Instead of poisoning the parasites, they endow crops with one or more of the parasites' genes, in effect immunizing them." As early as 1986, plant scientist Roger Beachy had inoculated tomato plants with a vaccine against the lethal tobacco mosaic virus. Beachy's genetically modified plants became completely immune to infection. Said Beachy in a recent interview, "What's been amazing to many of us is that we've seen advances that even were beyond our wildest expectations. We all knew it was theoretically possible, but to actually do it and employ it into the field. And then, at the end of four or five years, report that this has an advantage of increasing yields and reduce the use of agriculture chemicals by 50 million pounds [22.7 million kg] a year. It's an astounding number." Beachy's comments suggest that plant vaccines may be the best method of all in making PIPs.

Biotechnology laboratories currently make PIPs by the following steps:

1. Isolate a section of an organism's DNA containing the desired gene.
2. Add a *marker gene* to this DNA for use in step 6.
3. Treat a mixture of microbes so they can accept DNA into their cells (example: calcium chloride treatment to make the cell membrane permeable to DNA).
4. Mix the DNA containing the desired gene with the microbe.
5. Incubate the microbe-DNA mixture.
6. Identify the microbial cells that have absorbed the DNA by looking for traits of the marker gene, for instance, different color of microbial colonies or unusual reaction to antibiotics.

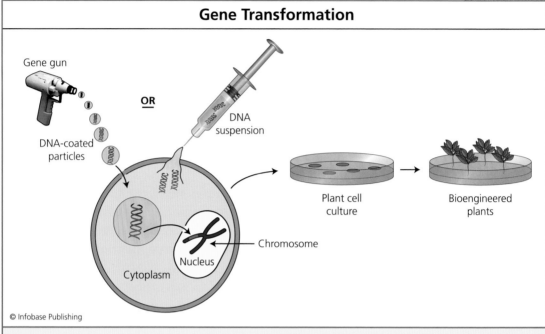

Gene Transformation

Gene gun

OR

DNA-coated particles

DNA suspension

Cytoplasm

Nucleus

Chromosome

Plant cell culture

Bioengineered plants

© Infobase Publishing

Green biotechnology puts plant-incorporated protectants (PIPs) into crops in a process called gene transformation. Gene transformation involves insertion of a gene into the DNA of another organism. Biologists physically force the new gene into plant cells by either injecting DNA or using a gene gun. Plant suppliers then sell either the transformed seeds or the seedling plants to growers.

7. Select only colonies that have the marker gene, meaning they also contain the new desired gene.

8. Isolate the new DNA from the microbes.

9. Put the new DNA into plant cells using either infective viruses or bacteria or by using a mechanical device called a gene gun.

The procedure of inserting DNA from one organism into another organism is called *transformation*; the new microbial or plant cells are transformed cells. Laboratory scientists grow the individual plant cells in a manner similar to growing microbial cells. After plant shoots emerge, technicians transfer the shoots to nutrients that induce roots to grow. After two to three weeks, the tiny plants have root systems strong enough to support the plants in soil. The bioengineered plants then grow strong in a nursery before workers plant them in fields or recover the seeds. Several details accompany each of these steps, of course, but this general scheme leads to today's bioengineered crops used in agriculture.

ENGINEERED FOODS AND CROPS

U.S. bioengineered crops consist mainly of soybeans, cotton, and corn; of the total amount of soybeans grown in the United States, about 85 percent is an engineered variety. Worldwide, corn, soybean, canola, and cotton lead all other bioengineered crops. Engineered varieties of cotton make up about 75 percent of all cotton, and engineered varieties of corn make up about 50 percent of all corn production. Tomato, potato, and tobacco growers also depend on bioengineered varieties to a lesser extent. Bioengineered crops cover at least 5 percent of the total land areas in both the United States and China, and that total may increase in the near future. The global increase in bioengineered crops will probably take place in countries that already pursue green biotechnology: Argentina, Brazil, Canada, India, Paraguay, and South Africa. The following table summarizes the major bioengineered food and nonfood crops growing today in U.S. agricultural fields.

BIOENGINEERED FOOD AND NONFOOD CROPS IN THE UNITED STATES	
TRAIT	**CROP**
insect resistant	corn, cotton, potato, tomato
virus resistant	potato, papaya, squash, watermelon, zucchini
mold resistant	peas
herbicide resistant	beet, corn, cotton, flax, rice, soybean
freeze tolerant	barley, Bermuda grass, canola
drought tolerant	cassava, corn, cotton, lettuce, rice

Consumers in the United States and Europe have been leery of the unknowns connected with bioengineered foods. Dr. Hugh Sampson of New York's Mount Sinai School of Medicine told *Newsweek* in 2001, "If a gene is moved from a common allergen like a peanut into something like a tomato, where no one expects it to be, it's a potential threat." In the United States, the U.S. Food and Drug Administration (FDA) requires only that bioengineered foods be labeled if they cause allergies or are significantly different from the natural food; European food producers must label all bioengineered items. As a consequence, consumers in Europe receive more information on bioengineered foods than Americans.

Quite a few restaurants have refused to serve bioengineered foods of any type because of a perceived health threat. The following safety issues have been put forth by concerned scientists and nonscientists regarding bioengineered crops:

- plant transgene (the foreign gene) spreading into ecosystems via pollen
- bioengineered animals mixing with native animals in an ecosystem
- pests overcoming pest resistance
- monocultures of bioengineered crops opening an opportunity for plant epidemics
- contamination of non-bioengineered foods with bioengineered variety

The biotechnology industry has tried to make the events listed here improbable, yet in biology improbable things sometimes happen. In 2000 a major food company recalled corn taco shells that had been contaminated with bioengineered corn that had not been approved by the FDA for human consumption. Joseph Mendelson, legal spokesperson for the food safety advocacy group Genetically Engineered Food Alert, blamed the FDA for the accident: "I view it as a very poignant cautionary tale that our regulatory system is not up to the task of preventing potential problems with genetically engineered food." Unfortunately, overworked FDA inspectors cannot be everywhere to prevent accidents.

This plant tissue laboratory at King Saud University in Saudi Arabia has the capability to grow plant tissue in small vessels inside incubators. Students use this to study the following disciplines in plant science: genetics, cell physiology, plant physiology, nutrition, weed biology, and infection. Green biotechnology also relies on plant tissue culture methods in genetic engineering. *(King Saud University)*

The total global acreage of bioengineered crops has grown since about 1996, chiefly because these crops afford farmers better yields and favorable economics. The issue of whether bioengineered crops injure or help the environment may be more difficult to determine. Increased yields and growing efficiency plus a decreased need for chemical pesticides certainly can help the environment. Better yield per acre also lessens the need of subsistence farmers to cut down *old-growth forests* for agriculture. Forest destruction has been occurring in many parts of Africa and Central and South America to the great detriment of ecosystems there. Additional attributes of bioengineered crops are discussed in the following "Case Study: Bangladesh Rice Research Institute."

Opponents of bioengineered crops contend that accidents such as the taco shell event can threaten ecosystems and possibly the health of people who eat bioengineered foods. Consumers have been particularly worried about unusual proteins in bioengineered food that could cause severe allergic reactions, as Dr. Sampson described. In 2000 the

Sustainable agriculture uses both biotechnology and organic farming, but many proponents of organically grown foods reject all bioengineering in agriculture. Organic foods help the environment because consumers can be assured that no chemical fertilizers, pesticides, herbicides, or meat-growth enhancers have been used and so cannot pollute the environment. *(Whole Foods)*

FDA commissioner Jane Henney told *FDA Consumer* magazine, "I understand why people are concerned about food allergies. If one is allergic to food, it needs to be rigorously avoided. Further, we don't want to create further new allergy problems with food developed from either traditional or biotech means. It is important to know that bioengineering does not make a food inherently different from conventionally produced food. And the technology doesn't make the food more likely to cause allergies." Though people remain skeptical of such claims, Henney explained, "All of the proteins that have been placed into foods through the tools of biotechnology that are on the market are nontoxic, rapidly digestible, and do not have the characteristics of proteins known to cause allergies." Thus far, bioengineered foods have an excellent safety record.

FUTURE AGRICULTURAL BIOTECHNOLOGY

Improved crop yield and quality have been the main goals of agricultural biotechnology thus far. Scientists have simultaneously worked on means for ensuring that biotechnology products do not cause harm if they escape into the environment. The following list shows some of the main investigations at U.S. universities and research centers and in other countries:

- feeds for improved yields in meat animals to avoid use of hormones and antibiotics
- drugs from milk-producing animals
- fiber-producing plants
- polymer- and plastic-producing plants

CASE STUDY: BANGLADESH RICE RESEARCH INSTITUTE

The Bangladesh Rice Research Institute (BRRI) operates nine regional stations that study improvements for Bangladesh's rice production. Bangladesh is a country on the Indian Ocean in which 95.5 percent of its land is bordered by India; Myanmar accounts for the remaining border. Bangladesh suffers from overpopulation, and the country's people live and work on land that is flood-prone or polluted. Shortages of clean water, forest loss, pollution from overuse of pesticides, and soil erosion have stressed the people of Bangladesh and worsened health conditions. BRRI now takes steps independent of the government to help ease food shortages. The organization's main focus resides in developing better rice varieties.

BRRI currently operates about 150 individual projects in rice research, either centered on farm management or technology for plant improvement. BRRI works on crop yield improvement in three main emphasis areas: (1) breeding, (2) hybrid varieties, and (3) biotechnology. Breeding programs make up BRRI's largest research group, with 20 different topics including insect resistance, disease resistance, salt tolerance, flood tolerance, cold tolerance, super high yield, and growth in non-salt tidal soil. The biotechnology projects at the BRRI regional stations are as follows:

⊙ methods for rice transformation

⊙ development of marker genes

⊙ genetic studies for grain or seed improvement

BRRI also acts as an agriculture extension service to help local growers combat diseases, pests, and other threats to their crops. Several rice farms have adopted integrated pest management methods that combine chemicals, biotechnology, and mechanized methods to increase yields.

Bangladesh is in the early stages of bioengineering experiments with Bt and gene transfer into rice plants. Though the government encourages

(continues)

(continued)

progress, many science students have yet to learn the techniques for working with DNA and nurturing plant cells in a laboratory, and so projects await trained biotechnologists to carry the work forward.

BRRI is unique because it stresses the need for safety assurances. Other parts of the world that confront starvation emphasize food supply over the safety concerns that dominate discussions about bioengineering in the United States. BRRI's greatest challenge hinges on the ability of biotechnology to make new products inexpensive to produce and to buy. Inexpensive crops produce less income for small farmers, however, so biotechnology must combine forces with agriculture policymakers if this technology is to rescue countries such as Bangladesh.

The Bangladesh Ministry of Science clearly sees a role for government to participate in attaining biotechnology goals. In its 2004 *National Biotechnology Policy* the ministry wrote, "The last part of the 20th century has witnessed spectacular progress in the fields of biotechnology and information and communication technology. Such advances have had a beneficial impact on food and health security. The tools and techniques of biotechnology will be used for poverty alleviation, health, nutrition and livelihood improvement, and conservation of environment." Though Bangladesh has obstacles ahead to overcome, it seems to be fully committed to using biotechnology to improve the circumstances of its people.

- drug-producing plants to avoid harvesting from old-growth forests
- food with enhanced nutritional quality
- nitrogen-absorbing plants to reduce needs for fertilizer
- plants that adapt to climate change
- plant and microbe removal of pollutants from soil and water
- improved drought/flood tolerance in plants
- high-salt tolerance in plants
- plants and trees with multiple pest/disease resistance

- improved plant seed strength or flower size
- improved plant and tree root distribution and absorbency
- improved fiber and wood production in fast-growing trees
- use of male-sterile plants or buffer zones to reduce the spread of pollen from bioengineered plants

CONCLUSION

Sustainable agriculture represents an attempt to make food production less energy-consuming and wasteful. To do this, proponents of sustainable agriculture have made use of biotechnology for substances that increase crop yield, make crops hardier, or help crops fend off infection from microbes, insects, or other parasites.

A sustainable farm is an operation in which food grows with the aid of natural pest control whenever possible, no tillage, and water and soil conservation. Sustainable farming went through a period of criticism when a few farmers first proposed it, but it has been steadily gaining ground in agriculture. Sustainable farming offers several approaches for farms, from 100 percent organic operations to methods that integrate chemicals with biological methods.

Biological methods in sustainable farming include biopesticides. Some biopesticides come from natural activities, such as one type of insect eating another more harmful type of insect. Other biopesticides come from laboratory methods in which scientists put genes into new plants or microbes. These biopesticides offer a variety of advantages, especially in decreasing the need for toxic chemicals, but they also create concern in the minds of many people regarding safety. Many opponents of bioengineering simply do not believe the risks outweigh the benefits.

The future of sustainable agriculture, agriculture biotechnology, and biopesticides rests with the ability of scientists to ensure the safety of consumers and the environment. Safety controls already appear in new products—suicide genes offer the most promise. But biotechnology has a responsibility to improve on all safety measures so that bioengineered organisms do not cause the hazards that many people fear. Science is never fail-safe, so farmers and large agricultural companies also must use new bioengineered products responsibly.

Sustainable agriculture's most important contribution will come from its capacity to meet two objectives simultaneously: It must increase crop yields while decreasing environmental harm that results from conventional farming. Fortunately, farms in the United States and abroad have demonstrated that these two objectives are possible. If carried out properly, sustainable agriculture protects ecosystems, conserves natural resources, aids biodiversity, and helps solve the world's hunger crisis. Sustainable agriculture certainly cannot solve all these problems by itself, but it can contribute in combination with other sustainable technologies.

WHITE BIOTECHNOLOGY

White biotechnology consists of biology-based techniques for helping industries make products that do not harm the environment. In different parts of the world, it may be referred to as industrial biotechnology, sustainable chemistry, or green chemistry. White biotechnology differs from green biotechnology in one major way: Green biotechnology solves individual problems in the environment, such as cleaning up mining wastes, but white biotechnology encompasses biological sciences and industrial processes for a more global effect on the environment. Certainly, white biotechnology adopts ways to clean up environmental damage, but it also seeks to achieve three specific goals: pollution prevention, resource conservation, and cost reduction. In other words, white biotechnology endeavors to protect the environment by taking preventive actions.

Green biotechnology results in the development of new forms of living things by transferring genes between dissimilar organisms. In contrast, white biotechnology makes use of living things to manufacture nonliving products. The chemical industry uses white biotechnology when it employs living organisms to make new chemicals, plastics, or strong long-chain molecules called polymers. Other industries have also taken tentative steps into white biotechnology: textiles, paper products, cosmetics, and food.

The main areas where white biotechnology has been used in the United States are the following: enzymes to replace chemical reactions in manufacturing processes; enzymes to replace synthesis steps that require high temperatures and high pressures; production-line sanitation using enzymes rather than chemical cleaners; microbes for water purification; and microbes for wastewater cleanup.

White biotechnology is called industrial biotechnology because it focuses on ways to make industrial processes more sustainable. That is, white biotechnology seeks to reduce industry's ecological footprint. Industry has been a major cause of environmental damage due to the pollution and wastes it generates and the natural resources it consumes. Industrialization plays a large part in global pollution and increasing greenhouse gases. These problems have prompted industries—some industries but certainly not all—to change the ways they operate. In many cases, these industries have created hazardous waste sites now controlled by the U.S. Environmental Protection Agency (EPA)'s Superfund program. White biotechnology might someday make most hazardous wastes a thing of the past.

The most formidable challenge in converting traditional manufacturing to white methods is that it requires an entirely new way of setting up production processes. White biotechnology depends not on strong chemicals or heavy machinery in its manufacturing steps but rather on nature's enzymes. The enzyme-based inventions that come from white biotechnology will probably look, sound, and possibly smell very different from the massive industrial activities that people have become accustomed to seeing. This chapter discusses the status of white biotechnology, one of the newest avenues toward sustainability in industry. It discusses the link between human activities and the ecological footprint and details the basics of calculating that footprint in "Case Study: The Science of Measuring the Ecological Footprint." Chapter 4 also covers the social and economic benefits of white biotechnology and the progress that other countries have made in adopting white biotechnology's methods.

BIOTECHNOLOGY AND INDUSTRY

Biological methods contribute only 5 percent of the chemical products manufactured in the United States, but that is expected to increase to between 10 and 20 percent by 2010, according to the publication *White Biotechnology: Gateway to a More Sustainable Future* (2003) by the consultants McKinsey and Company. Biological methods are not new to industrial processes. Water-purification facilities have for a long time used bacteria to remove organic contaminants from water to make it safe to drink. Microbes have been used since antiquity to produce or preserve foods, and microbial enzymes have been harnessed for a variety of jobs, from tanning leather to giving detergents stain-fighting ability. Makers of silk, wine, or chemical alcohols such as ethanol use natural organisms to make their products.

When the chemical industry began its rapid growth period in the 1940s, the conveniences it produced seemed like valuable assets to society. But chemical companies also caused pollution and depleted natural resources. People who looked beyond the immediate future probably realized that industries could not continue on that path, yet today many companies promise infinite growth to their stockholders even though the world holds only finite resources. The next 50 years will determine whether companies continue growing into behemoths or if they will be forced to downsize and adopt sustainable methods. Biological processes provide several advantages in manufacturing that conventional methods cannot offer, as follows:

- Biology produces biodegradable products and wastes.
- Biological reactions require less energy to run than mechanized methods.
- Biological reactions tend not to pollute water, soil, or air with hazardous compounds.
- Biological reactions are quiet, so do not contribute to noise pollution.
- Biotechnology holds promise in finding alternatives to fossil fuels.

If government regulations on hazardous materials become stronger in the future, industries may turn to white biotechnology sooner rather than later, but some industry leaders might be reluctant to convert their operations to white methods because they fear high costs. Any new technology has high costs initially until more people begin to use it. As more companies adopt sustainable methods or biological methods, the costs of these choices decline. Molecular biologist Giovanni Frazzetto wrote in a 2003 newsletter published by the European Molecular Biology Organization, "A major limitation of the commercialization of . . . bacterial plastics has always been their cost, as they are 5–10 times more expensive to produce than petroleum-based polymers. Much effort has

White Biotechnology and Manufacturing

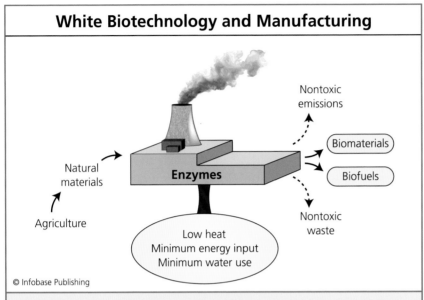

Nontoxic
emissions

Biomaterials

Biofuels

Natural
materials

Enzymes

Agriculture

Nontoxic
waste

Low heat
Minimum energy input
Minimum water use

© Infobase Publishing

Industrialized countries, including the United States, have been slow to make use of white biotechnology. Changing from conventional manufacturing to white methods involves a switch from high-energy, high-heat systems that depend on chemicals to low-energy, moderate-heat systems that depend on enzymes. White biotechnology decreases energy use and hazardous wastes that come from current manufacturing methods. National governments and international groups may need to encourage more industries to make this critical transition.

therefore gone into reducing production costs through the development of better bacterial strains." Frazzetto went on to point out that plant-made plastics may eventually prove to be more economical than microbial plastics.

White technologies require a blend of disciplines to give industry its best chance of changing from conventional manufacturing to biology-based manufacturing. In addition to the use of microbes, plants, or the enzymes extracted from plants, white biotechnology will need to combine chemistry, engineering, and computer science. Chemists will work with biotechnologists to find new biological ways to run classic chemical reactions. The manufacturing plants will need engineers to design new types of production lines, and computer scientists will build systems that allow the entire process to run efficiently. At the same time, white biotechnology must call upon genetics, molecular biology, microbiology, and plant science to develop each new generation of products.

CASE STUDY: THE SCIENCE OF MEASURING THE ECOLOGICAL FOOTPRINT

The Earth's human population has an ecological footprint 23 percent greater than the world can replenish. Viewed another way, the Earth needs about 450 days to replenish the things people use up in 365 days. How do humans manage to endure even though they are depleting the planet's resources?

The 23 percent of the Earth's capacity that people have exceeded is called ecological overshoot or, simply, overshoot. Overshoot varies over the face of the globe; some regions do not exceed their ecological footprints, and other regions overshoot their resources by a wide margin. The world's total population manages overshoot in one of three ways: (1) regions that overshoot their resources use resources from other parts of the world that have excess; (2) regions use new technologies to help spare their resources; or (3) regions keep using up their resources so that future generations will not have enough to sustain a similar lifestyle. This third tactic is leading to a serious global predicament that would hardly be the best choice for managing overshoot.

The concept of the ecological footprint gives scientists an idea of the sustainability of lifestyles on a continent, country, city, or small town. Sustainability relates to activities that conserve and replenish resources at the same rate in which the resources are used. Calculating footprints has helped show which types of lifestyles eat up the planet's reserves and which lifestyles conserve resources in a sustainable fashion. Ecological footprints of different types of lifestyles—rural, agrarian, urban, industrialized, service-oriented—follow a general pattern: Industrialized regions often overshoot their resources, and agrarian regions maintain an excess of resources.

Online footprint calculators allow people to enter data that describe their lifestyles and then determine the ecological footprint. The main categories of information that must be entered into the calculator are the following:

1. continent
2. size of house and how many people living in it
3. type of vehicle and average daily driving distance

(continues)

(continued)

4. type of energy usage and general amount used

5. type of diet

6. amount of recyclable and nonrecyclable wastes generated

The resulting ecological footprint can be measured in global acres, in global hectares, or in planets. A measurement described as a planet signifies the number of Earths it would take to support humanity if everyone lived exactly according to the values a person has entered into the footprint calculator. The Earth's current human population uses 1.3 planets per year, clearly an overshoot situation. Some calculators provide a more detailed footprint by figuring the amount of the following resources a person uses up every year: energy-producing land, cropland, grazing land, forested land, developed land, and fishing grounds.

The following table provides the names of organizations that offer easy-to-use ecological footprint calculators on their Web sites.

Ecological footprints vary from person to person. For instance, a businessperson living in Connecticut and commuting daily and alone in a large car to New York City, taking frequent flights, and occupying a spacious house in the suburbs has a much larger footprint than a farmer who grows most of his own foods and lives in a modest home in Montana powered by renewable energy.

People can lower their ecological footprints by adjusting various activities in an almost limitless assortment of resource-conserving choices, as follows:

- energy conservation in a small home
- renewable energy sources
- locally grown foods and limited amounts of fish and meat
- short distances and infrequent driving
- car-sharing or commuting by walking or bicycling
- use of public transportation
- very limited or no air travel

SUSTAINABLE CHEMISTRY

Sustainable chemistry consists of chemical reactions and chemical manufacturing using safer, more efficient, and more environmentally favorable methods than are used in traditional manufacturing. Several industries

- preferred use of recyclable packaged products
- maximum recycling of paper and plastics

Most Americans achieve some of these actions but not all of them. That explains why in a 2006 *Washington Post* article, the reporter Bridget Bentz Sizer said that the average American lifestyle requires 5.3 planets to sustain. Achieving a sustainable lifestyle is hard work and requires thoughtful decisions. As ecological footprints often show, sustainability is overdue.

ECOLOGICAL FOOTPRINT CALCULATORS		
ORGANIZATION	FEATURES OF THE CALCULATOR	REGIONS COVERED
Best Foot Forward	footprint calculated as number of planet Earths; simultaneously calculates carbon footprint and carbon dioxide (CO_2) emissions	United Kingdom
Earth Day Network	footprint calculated as number of planet Earths; allows user to create personal profile	United States; Australia
Ecology Fund	footprint as hectares, acres, or planets; quick, one-page questionnaire	United States; Europe; Australia
Global Footprint Network	footprint calculated as number of planet Earths; allows user to create personal profile; includes basics on carbon footprint	United States; Australia
The Green Office	footprint calculated as acres for businesses	United States
World Wildlife Fund	footprint calculated as number of planet Earths	United Kingdom

depend on chemical reactions to convert a raw material into a substance that consumers need, and very often these reactions require a high input of energy or produce a large amount of dangerous by-products. The following table describes a list of industrial processes that have been dependent on chemical reactions—some of these methods date back centuries with little change to the present day—and the new white biotechnology methods that replace them. Appendix B provides a list of all the global industries that can take advantage of biotechnology to improve their sustainability and efficiency. As efficiency of industrial processes increases, the costs of running those processes almost always decrease.

Chemical companies have the best opportunity to take advantage of white biotechnology, especially for making plastics and other polymers or diesel and other fuels. Petroleum products often serve as the raw material, called a *feedstock,* to start the process of making either a plastic or a fuel. Petroleum chemicals supply the building-block molecules that, when strung together, make long, chainlike polymers. These polymers then serve as the feedstock for plastic products, polyesters, and thermoplastics, which are plastics that soften when heated and harden when cooled. White biotechnology bypasses this need for petroleum by using natural chemical reactions to convert grain, usually corn, to polymer molecules by the following steps.

Corn sugar → lactic acid → polylactic acid → polymer → plastics

The goal of the biotechnology industry is to soon branch out into using feeds other than corn to achieve the same result. The Biotechnology Industry Organization (BIO) has estimated that the United States alone produces 175 billion pounds (79 billion kg) of organic chemicals each year, which means it is using up an enormous amount of corn. Finding sources in addition to corn for polymer production would help farmers and strengthen agriculture's business future.

One of sustainable chemistry's biggest challenges resides in finding new feedstocks to replace fossil-fuel feedstocks. This challenge needs to be overcome as quickly as possible for two reasons. First, petroleum-based fuels cause a significant amount of environmental pollution, and their by-products in vehicle exhaust contribute to global warming. Second, economies worldwide, notably in the United States, react to the availability, price, and potential future supply of petroleum. White biotechnology must find ways not only to help nations move from dependence on petroleum fuels, but also to make this transition a smooth one with little disruption to national economies. Though this may seem an almost impossible task, the alternative fuels industry has made good progress in promoting new types of fuels.

Biofuel is the name for the most well-known alternative fuel. Biofuel production occurs when natural enzymes convert cellulose fibers in grains

Uses of White Biotechnology

Product	Old Manufacturing Process	White Biotechnology Process	Technology Used	Consumer Benefits
bleached paper	wood chips boiled in chemicals and then bleached to make pulp	enzymes degrade wood fibers and cell walls	bioengineered microbes to produce bleaching enzymes	saves energy; releases fewer toxic dioxin pollutants
detergents	phosphates added as brightening and cleaning agents	enzymes remove stains and brighten fabrics	bioengineered microbes	saves energy; eliminates phosphate water pollution
ethanol fuel	grains fermented into ethanol	cellulase enzyme converts plant fibers to sugar, which converts to ethanol	bioengineered microbes to produce cellulase and ferment sugars to ethanol	alternative to fossil fuels; less greenhouse gas emissions
polyester	chemically produced from petroleum by-products	polyester produced biologically from corn starch	microbial fermentation of corn sugar to a polymer that converts to polyester when heated	biodegradable; nontoxic constituents; avoids fossil fuel use
vitamins and antibiotics	complex chemical synthesis steps	one-step fermentations	bioengineered microbes	saves energy; saves costs; reduces chemical wastes

The biotechnology industry has produced more than 500 drugs and medical devices. Biotechnology also develops substances for cleaning up pollution, replacing agricultural chemicals, and enhancing the characteristics of food crops. About 1,500 biotechnology companies operate in the United States, which has more than half of all biotechnology companies worldwide.

to ethanol; when this is done by biological means, the product is called *bioethanol*. BIO's 2004 report "New Biotech Tools for a Cleaner Environment" stated, "According to Professor Bruce Dale of Michigan State University, bioethanol from cellulose generates eight to ten times as much net energy as is required for its production. It is estimated that one gallon [3.8 l] of cellulosic ethanol can replace thirty gallons [113.5 l] of imported oil equivalents." In this context, net energy equals the energy left over after expending some of the fuel's total energy in its production; an oil equivalent is a unit of petroleum oil having the same energy content as bioethanol.

BIO has further proposed that industry need not rely on corn to make bioethanol. In fact, reliance on corn for making alternative fuels has created a few problems of its own. Corn-based fuel requires energy for cultivating, storing, transporting, and producing, many of the steps of which rely on gas-guzzling trucks. The reporter Jeffrey Stinson of *USA Today* pointed out in 2008, "The use of corn and sugar to make ethanol is a main driver of rampant inflation in worldwide food costs during the past year.

Grocery bills are up across Europe, and the United Nations World Food Program says that rising food prices have pushed 100 million people into hunger worldwide." The energy inefficiencies of corn-based fuel and its impact on world food prices, therefore, take some of the luster from the once-promising idea of using corn for energy.

Industrial chemists would be wise to find bioethanol sources in addition to corn because of corn's drawbacks. Residues left in the field in no-tillage farming or sludge residues from paper production would both serve as good raw materials for bioethanol production. Almost all the ideas that have been proposed recently for finding new feedstocks, improving energy efficiency, and reducing costs depend on the main attribute of biological reactions: enzyme activity. Enzymes carry out activities with efficiency and safety that chemical reactions cannot match. The following sidebar, "Enzyme Kinetics," describes how scientists in white biotechnology use the attributes of natural enzymes to create new products that serve the needs of people and preserve the environment.

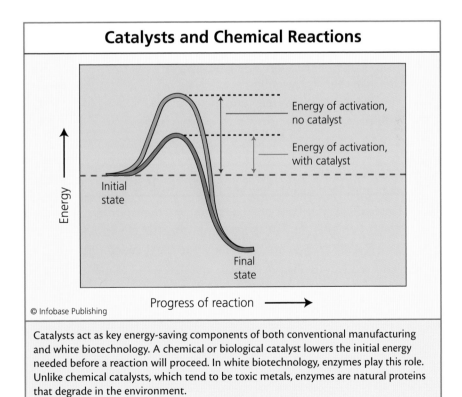

Catalysts and Chemical Reactions

Energy of activation, no catalyst

Energy of activation, with catalyst

Energy

Initial state

Final state

Progress of reaction

© Infobase Publishing

Catalysts act as key energy-saving components of both conventional manufacturing and white biotechnology. A chemical or biological catalyst lowers the initial energy needed before a reaction will proceed. In white biotechnology, enzymes play this role. Unlike chemical catalysts, which tend to be toxic metals, enzymes are natural proteins that degrade in the environment.

ENZYME KINETICS

Biotechnology relies on a process called biocatalysis to carry out the steps leading to new plastics, fuels, drugs, or any other biotechnology product. A catalyst is a substance that speeds up a reaction without being used up in that reaction. This process of changing a compound from one form to another form is *kinetics,* the study of conditions that are in a constant state of change. White biotechnology relies on enzyme kinetics to make products in an entirely new way compared with traditional manufacturing.

Catalysis is an important part of enzyme kinetics because a catalyst makes a reaction more energy efficient than it would be without the catalyst. This chemically favorable condition is like a person riding a skateboard toward a hill. If the skateboard does not contain enough energy, it will not glide up and over the hill to the other side. A catalyst works not by making the skateboard go faster so it can get over the hill but, rather, a catalyst lowers the hill.

In biochemistry, enzymes serve the same purpose as chemical catalysts—the reason they are called biocatalysts—by lowering the energy a reaction requires to go forward. (Biocatalysis usually refers to the synthesis of organic chemicals.) If a student were to draw a graph of the hill, the height of the hill from the ground to the top is called the *free energy of activation.* Enzymes make many manufacturing processes more efficient and less energy consuming by lowering the free energy of activation. Enzymes also provide an added benefit of producing less hazardous by-products than chemical reactions.

White biotechnology studies the kinetics of the following five main classes of enzymes:

SOCIAL AND ECONOMIC BENEFITS

White biotechnology delivers obvious benefits to the environment, as discussed in this chapter. If industries commit to the new manufacturing processes offered by white biotechnology, they will create other benefits such as cost reductions, energy savings, and decreased amounts of waste. White biotechnology may provide additional secondary benefits to society and business. In fact, white biotechnology is one of the few initiatives that ties society's needs to business needs, that is, the need to decrease the ecological footprint while earning a profit.

Society has paid a price for relying on fossil fuels as its main energy source. Air, soil, and water pollution have resulted from the transport and burning of fossil fuels, and international conflicts have arisen over

- proteases that degrade proteins
- lipases that break down fats
- amylases that digest starch
- cellulases that break apart the plant fiber cellulose
- specialized enzymes that carry out specific reactions or produce specific bio-products, which are products made partially or wholly by biological reactions

Polymerization enzymes provide an example of specialized enzymes. Plastics and polymer man-ufacturers use polymerization enzymes to connect monomer units in single file to build long, chainlike polymer compounds.

The Danish company Novozymes has become a leader in white biotechnology by designing specialized enzymes for industrial and home uses. The company touted its products in 2008: "Probably the biggest advantage of enzymes in organic synthesis is their remarkable chemical selectivity. This gives a number of commercial benefits such as better production of [the correct form of a desired product], fewer side reactions, easier separation of products and less pollution; all of which adds up to lower costs." Though Novozymes certainly wants to stress the value of its enzymes, the company's statement accurately sums up the advantages of using enzymes in manufacturing.

petroleum. But the rush to develop biofuels from corn may be exerting its own pressures on small farmers, poor countries, and the environment. Massive conversion of farmland from a variety of crops to corn causes groundwater depletion, soil erosion, and fertilizer-polluted runoff. More troubling, the growing emphasis on biofuels raises food prices, which threatens impoverished areas. Biofuel policy adviser Rob Bailey shared his thoughts on biofuel production with *BBC News* in 2008: "If the fuel value for a crop exceeds its food value, then it will be used for fuel instead." Bailey then added his viewpoint on this subject. "Rich countries . . . are making climate change worse, not better. They are stealing crops and land away from food production, and they are destroying millions of liveli-hoods in the process." Of course, a contrasting argument could be that increased demand for biofuel crops gives the poor an opportunity for earning an income. Many of the poorest and food-deprived regions of the world, however, lie on land that has been destroyed by drought or pollution or other environmental ills.

Agricultural biotechnology likely can improve conditions in poor developing countries by increasing crop yields in parts of the world that depend on farming for survival. For this reason, agricultural biotechnology and white biotechnology might work together to solve problems associated with global food production. Can white biotechnology become a long-term solution to improve economies and lift people out of poverty and hunger? Feika Sijbesma of the Dutch company Royal DSM offered this vision in a 2007 presentation in Lyon, France, on biotechnology, poverty, and hunger: "White biotechnology is rapidly gaining momentum as a cost- and eco-efficient technology to produce bio-based fuels, chemicals, materials and specialties in a safe and sustainable way from renewable resources using nature's toolset." White biotechnology admittedly has distance to cover before making a meaningful impact on the environment, helping to solve society's biggest problems, and all the while making economic sense to industries.

PRESENT AND FUTURE WHITE BIOTECHNOLOGY

Today, white biotechnology accounts for less than 10 percent of the industries that can make products biologically. In Europe, industries have expressed interest in white biotechnology but only if assured of financial advantages. The United States has taken steps to implement white biotechnology, but when considering all U.S. industries, white biotechnology has not yet become a standard way of doing business.

In 2000, university scientists and industrial representatives held a conference called Vision 2020 to discuss the chemistry profession's present and future. Attendees discussed the new chemistries and engineering techniques likely to emerge in the next two decades. Then as now, innovative polymers were identified as one specialty expected to grow in the near future, but at that time few companies had taken steps to convert their traditional production methods to biological methods. White biotechnology remains more a vision than a reality in U.S. and European industries, especially for companies that cannot make use of enzymes.

China has been building a white biotechnology industry for the past several years. "China is a leading producer of many vitamins and polysaccharides," reported a business press release in 2008. "Many companies in China intend to utilize lignocellulosic raw materials to produce fuel ethanol, with pilot-scale plants . . . being built up in China." (Lignocellulosic materials are strong, hard-to-degrade fibers made of plant structural substances lignin and cellulose.) Because China is at the upturn of its industrial revo-

White biotechnology includes the use and the production of biofuels. Westchester County, New York, operates this Veggie Van, which runs on fuel made from the 10,000 gallons (37,843 l) of used vegetable oil produced by county-owned concessions each year. *(Westchestergov.com)*

lution, it has a valuable opportunity to incorporate white biotechnology, especially when considering China's serious pollution problems.

The sidebar "Organisation for Economic Co-operation and Development (OECD)" describes one international organization that acts as a resource for sciences such as white biotechnology. Several national and international programs like OECD pursue all feasible technologies for reducing greenhouse gases and finding alternative energy sources.

Academia and industry will need to work together to make these goals grow within reach. For academia, present and future programs target the following three research areas: (1) the impact of new chemicals on human health and the environment; (2) cutting-edge research for developing technologies for use by industry; and (3) training a new generation of chemists, biochemists, and biologists to make white biotechnology a reality. Industry will likely take on its own responsibilities highlighted by the following: (1) developing and manufacturing materials and products that are safe for people and the environment; (2) continually improving manufacturing processes to save and reuse energy; and (3) providing the public with all the information they need to understand the new generation of products, their advantages and disadvantages.

In the United States, the EPA has published 12 principles of the future's cleaner chemistry on its Web site:

1. Prevent waste.

2. Design safer chemicals and products.

3. Design less hazardous chemical syntheses.

4. Use renewable feedstocks.

5. Use reusable catalysts rather than onetime use chemicals.

6. Avoid chemical derivatives that generate extra waste.

7. Design processes so that the product contains the maximum amount of starting materials (called maximizing atom economy).

8. Employ safer solvents and reaction conditions.

9. Increase energy efficiency.

10. Design degradable products and chemicals.

11. Monitor the processes as they occur to prevent pollution.

12. Minimize the potential for accidents.

ORGANISATION FOR ECONOMIC CO-OPERATION AND DEVELOPMENT (OECD)

The OECD is an international organization of about 30 countries that share knowledge regarding the world's economy. Member countries also cooperate in information-sharing in the following areas: (1) economy, (2) development, which seeks to aid developing countries' participation in the global economy, (3) governance, (4) environmental sustainability, (5) society, including education, health, employment, and social issues, (6) finance, and (7) technology innovation. Most important, this organization helps set policies on how to pay for the innovations developed in all of these research areas.

In 1998, OECD began an initiative in sustainable chemistry for the purpose of developing environmentally safe chemicals to replace existing hazardous chemicals. The organization expressed its rationale for pursuing sustainable chemistry as follows: "Our life is supported and enriched by an enormous variety of chemical products that have been invented by chemists and chemical engineers, and produced by the chemical industry. But at the same time, we have learned from experiences of the past decades that OECD countries are faced with environmental concerns including the unsustainable use of nonrenewable natural resources,

Transformation from conventional industry to sustainable industry certainly does not occur overnight. New technologies simply do not plug into old equipment. These technologies require laboratory testing, pilot-scale testing, scale-up to manufacturing levels, and deployment of the new equipment with several test runs. Plant managers will likely backtrack during the development phase to find and fix errors. Engineers meanwhile will redesign portions of the production line to avoid problems as they arise during scale-up. Throughout all these steps, employees must learn the techniques of using biological reactions rather than chemical reactions. The world needs industries to change to white biotechnology as soon as possible. The reality, however, hints that white biotechnology is a long-term project in environmental science.

INTERNATIONAL PROGRAMS

The EPA collaborates with international programs such as the OECD for developing and implementing sustainable technologies. These international

the degradation of ecosystems and the disruption of the environmental systems that support human life. The general public now expects chemists and chemical engineers to work to minimize and reduce effects on human health and the environment associated with the production and use of chemical products." Sustainable chemistry "can contribute to achieving a cleaner, healthier, and sustainable environment and improving the image of chemistry as a problem-solving science in society." In other words, the OECD considers the social aspects of technological innovations.

OECD member nations come principally from industrialized North America, Europe, and Asia. The member countries are: Australia, Austria, Belgium, Canada, the Czech Republic, Denmark, Finland, France, Germany, Greece, Hungary, Iceland, Ireland, Italy, Japan, Korea, Luxembourg, Mexico, the Netherlands, New Zealand, Norway, Poland, Portugal, the Slovak Republic, Spain, Sweden, Switzerland, Turkey, the United Kingdom, and the United States. By working in cooperation, these countries may soon help each other make meaningful advances in white biotechnology.

programs emphasize different aspects of white biotechnology, such as training, chemical research, pollution prevention, or introducing white biotechnology to developing nations. The following table presents the major organizations that have international interests in promoting and using white biotechnology.

New technologies for safer and less resource-demanding manufacturing will be easier to discover in the near future if countries pool their talents for building sustainability.

INTERNATIONAL ORGANIZATIONS PURSUING WHITE BIOTECHNOLOGY		
ORGANIZATION	COUNTRY	MAIN EMPHASIS
Royal Australian Chemical Institute	Australia	alternative synthetic pathways, reactions, and chemicals
Green Chemistry of Brazil (*Química Verde no Brazil*)	Brazil	pursuing projects related to the EPA's 12 green chemistry principles
Environment Canada	Canada	pollution prevention
Federal Ministry for the Environment, Nature Conservation and Nuclear Safety	Germany	advances in chemistry
Interuniversity Consortium of Chemistry for the Environment	Italy	university research in sustainable chemistry
Japan Chemical Innovation Institute	Japan	collaborative research—industry, government, academia—in sustainable chemistry
Royal Society of Chemistry	United Kingdom	funding for research in sustainable chemistry
American Chemical Society Green Chemistry Institute	United States	sustainable chemistry education

CONCLUSION

White biotechnology consists of industrial processes based on biological reactions rather than chemical reactions. The purpose of this technology is to save energy and reduce wastes and pollution yet continue to allow industries to be profitable. The ultimate goal of white biotechnology focuses on reducing humanity's ecological footprint by reducing industry's global footprint.

White biotechnology uses sustainable chemistry to meet its objectives. Sustainable chemistry involves the use of biological enzymes, moderate temperatures and pressures, and biodegradable end products. In contrast, conventional chemistry often relies on hazardous chemicals at high temperatures or pressures to drive synthesis reactions. At the same time, the traditional chemical industry produces large amounts of hazardous waste that persist in the environment for a long time.

The environmental benefits of white biotechnology ultimately lead to social benefits by decreasing pollution, lowering the emission of greenhouse gases, and slowing climate change. Poverty, drought, and population shifts relate to global climate change, so in an overall sense, white biotechnology may someday pull people out of poverty by making their environment healthier and able to support their numbers. White biotechnology is in its early stages, however, and has not yet given any meaningful benefits to the environment and society. Industry finds it difficult to change from systems that have worked for more than a century to plunge into the unknown world of reactions and enzymes extracted from microbes, plants, and animals. The future of white biotechnology therefore remains in question. As the human population continues to exceed its ecological footprint, however, white biotechnology should play a larger role in building sustainable industry practices.

MARINE BIOTECHNOLOGY

Marine biotechnology uses living organisms from marine habitats for industrial processes or for making consumer products. This so-called blue biotechnology has contributed to drug discoveries and the production of cosmetics, industrial raw materials, and nutritional supplements. The oceans cover a little more than 71 percent of the Earth's surface, and their waters contain unique habitats unlike anything on land. For instance, the ocean's life-forms have adapted to salt water, intense pressure, total darkness, frigid waters, or incredibly hot steam emissions.

Regardless of the extreme habitats in which they live, most marine organisms take part in food chains that begin with tiny creatures called *plankton*. Plankton are one-celled or multicellular organisms that supply nutrients to more complex marine organisms. Plankton of animal origin is called zooplankton, and plankton of plant origin is called phytoplankton; phytoplankton is additionally important because it captures the Sun's energy by photosynthesis.

Marine biotechnology focuses on two main aspects in environmental science: (1) discovering new substances that improve the environment on land and (2) inventing new biological processes to improve ocean ecosystems. The biggest breakthroughs in marine biotechnology today result from research that takes place at universities, independent marine science centers, and government marine science centers. Appendix C lists the major international marine institutes with first-rate programs in biotechnology.

Marine biotechnology can be categorized into the following focus areas:

- bioremediation of oil spills and other wastes on the seafloor and coastal areas
- bioremediation of salt marshes
- marine environmental health regarding parasites, toxins, and diseases
- diagnostic methods for assessing marine organism health
- coral reef and other ecosystem remediation
- marine conditions and human health
- enhancement of endangered marine species

Like green biotechnology, marine biotechnology depends on the discovery of useful genes in marine organisms that can be used in bioengineering. Rather than study each newly discovered gene one at a time, however, marine and other types of biotechnology have turned to genomics. Genomics is the study of how genes line up in sequence in deoxyribonucleic acid (DNA) or ribonucleic acid (RNA). DNA is a large molecule made of repeating sugar, nitrogen, and phosphate units in a doubled-stranded structure; DNA contains all the genes of an organism. RNA is similar but not identical to DNA, contains a single strand, and carries out steps in DNA replication and protein synthesis inside cells. The gene sequence in DNA contains all the information that describes a living organism. Marine scientists use genomics for studying three main topics: (1) how multiple genes have evolved; (2) how these genes function today; and (3) the ways in which these genes work together. Marine biotechnology includes both laboratory experiments in genomics and on-site studies on the ocean's surface, under the surface, or along coasts.

This chapter describes the ocean's marine life and the reasons why it is critical to the planet's health. It also describes the techniques scientists use to study ocean ecosystems, especially plankton, which may be the most important organisms to life on land and in the sea. Two main topics in marine science are reviewed: aquaculture and remediation. The chapter provides a discussion on the uses of both plant and animal aquaculture; it also details marine remediation methods, especially those used in coral reef restoration. As a general theme, chapter 5 explains the dual objectives of marine science: protecting oceans from ecological harm and studying the ocean as a natural resource.

THE WORLD'S MARINE LIFE

In the future, humans may learn to live without certain natural resources, such as oil, coal, or precious gems, should those resources run out. Humans and other organisms that live on land cannot, however, last long without the oceans and the marine life they contain. The world's oceans hold more life than any other place on Earth, and categorizing all the ecosystems that can be found in the ocean soon becomes a bewildering task. For ease in studying marine habitats, marine ecosystems usually belong to one of three distinct types: (1) the neritic zone, (2) the oceanic zone, or (3) the benthic zone. The following table summarizes the characteristics of these three marine ecosystem categories.

Marine biologists divide the three marine habitats further to give more detailed descriptions of the conditions found there. Different conventions for categorizing these zones have been used; one standard classification includes four zones based on light penetration. This classification system is shown in the following list. Depending on ocean conditions, the depths of these zones vary from season to season and in different parts of the world.

- coastal zone—extends from the high-tide mark to the continental shelf and contains warm, nutrient-rich, shallow water and photosynthetic organisms

TYPES OF MARINE ECOSYSTEMS		
NAME	LOCATION	DESCRIPTION
neritic zone	from the shore to a few miles from shore	shallow; warm; high levels of nutrients and phytoplankton; high biodiversity
oceanic zone	open ocean to about 600 feet (183 m) in depth	contains photic zone that receives abundant light to depths where no light penetrates (590 feet [180 m]); moderate biodiversity
benthic zone	continental slope to the ocean floor	very cold temperature; no light; constant downward flow of decaying organic matter

Marine life is the most diverse in the world, and many undiscovered species may produce substances that humans can use. Marine scientists study ecosystems that are difficult to reach and contain many species that have not yet been identified and have been altered by warming ocean temperatures. This coral reef is in the Gulf of Aqaba in the Red Sea. *(Mohammed Al Momany, NOAA/Department of Commerce)*

- euphotic zone—upper depths of the open ocean from the coastal zone seaward; receives ample sunlight, so contains photosynthesis
- bathyal zone—below the euphotic zone; receives low light; sometimes called the twilight zone
- abyssal zone—dark, deepest waters to the ocean floor; contains deep canyons or trenches

It has often been said that scientists know more about outer space than they know about the deepest parts of the ocean. Bruce Robison of the Monterey Bay Aquarium Research Institute told *National Geographic* in 2004, "The deep sea is the largest habitat on our planet, and yet it's still the most unexplored." For many years, marine scientists thought little life occurred on the ocean floor at depths from 6,500 feet to 33,000 feet (2,000–10,000 m), which is from one mile to more than six miles deep

(1.6–9.6 km). In 1987, marine biologists James Childress, Horst Felbeck, and George Somero explained in *Scientific American,* "Biologists categorize many of the world's environments as deserts; regions where the limited availability of some key factor, such as water, sunlight or an essential nutrient, places sharp constraints on the existence of living things. Until recently the deep sea was considered to be such a desert, where the low abundance of organisms stems from the extreme limitation of the food supply." As marine science technologies became more sophisticated, marine biologists discovered quite a different circumstance.

Automated deep-sea robots have since then helped in uncovering not a barren landscape, but a habitat teeming with unique life. Scientists have explored the ocean floor and have even sent probes down ocean canyons more than 12,000 feet (3,656 m) deep. In those places, they have discovered marine life that resembles nothing like the life found in shallower waters. "It's a world so unlike our own that we really shouldn't be surprised to find animals that defy our imagination," said Robison. The sidebar "Global Marine Protected Areas" explains why these remote regions are of such high value to science.

The marine life in the deep ocean probably subsists by using unique metabolic pathways run by enzymes that function in very cold temperatures (approaching 3°F; -16°C), under high pressure (approaching 16,000 pounds per square inch [psi]), and with little oxygen. The deep-sea marine life at about 2,300 feet (701 m) or deeper possesses another quality: bioluminescence. *National Geographic* writer Virginia Morell noted in the same 2004 article, "Nearly every creature we see at this depth—fish, squid, invertebrates—is bioluminescent, equipped with special organs called photophores that emit light via chemical reactions." Bioluminescence has been one of the first marine activities that scientists have tried to incorporate into new technologies on land.

Deep-ocean marine life may become a resource for new enzymes to carry out reactions that have been difficult and expensive to conduct in laboratories. The bioluminescence apparatus of these creatures seems a perfect component of *biosensors,* which are probes containing a biological component and a chemical component for detecting compounds in nature. None of these plans will work if the ocean loses its vitality due to pollution or overfishing. The following sidebar, "Global Marine Protected Areas," explains some of the approaches to protecting marine habitats.

GLOBAL MARINE PROTECTED AREAS

Marine protected areas (MPAs) consist of ocean regions that have some legal restrictions on their use. The reasons for these protections vary throughout the world, but usually have been implemented for one of the five following reasons: ecosystem protection; habitat protection for rare or overfished species; fish nursing grounds; natural resources conservation; or preservation of natural or cultural heritage. In the United States, protection of MPAs ranges from restricted recreational use to complete banning of all human presence.

United States and global MPA management contain a confusing variety of authorities and rules on how MPAs may be used. Sometimes the restrictions on different types of MPAs overlap, as the following table shows. In the United States, the National Marine Protected Areas Center in Maryland categorizes MPAs into groups based on the restriction's primary purpose, also described in the following table.

MARINE PROTECTED AREA CLASSIFICATIONS		
PURPOSE OF THE MPA OR PURPOSE OF THE PROTECTION	DESCRIPTION	U.S. EXAMPLES
conservation	conservation of biodiversity, deepwater habitat, or historical shipwrecks	marine sanctuaries, fisheries, wildlife refuges
ecology	conservation of ecosystem or habitat	marine sanctuaries, national parks, wildlife refuges
permanence of the protection	permanent protection, temporary protection, conditional (renewable) protection	permanent marine sanctuaries, temporary area closures to allow recovery of species, protected areas expected to become permanently protected (conditional)

(continues)

(continued)

MARINE PROTECTED AREA CLASSIFICATIONS
(continued)

PURPOSE OF THE MPA OR PURPOSE OF THE PROTECTION	DESCRIPTION	U.S. EXAMPLES
constancy of protection	year-round, seasonal, rotating	year-round marine sanctuaries and national parks, seasonal closures for habitat recovery, rotating area closures to reestablish fisheries
level of protection	varying levels of human use for protecting the area and its marine life	uniform multiple-use MPAs, zoned multiple-use MPAs, no-take zones

An MPA's major emphasis should be the protection of ecosystems and species; the best way to protect both is by protecting habitat. MPAs serve to ensure that rare or as-yet-undiscovered marine species do not disappear from the Earth due to habitat destruction. The impact of MPAs on conserving biodiversity is not fully known, so their importance to marine biotechnology has yet to be determined. Of course, extinct marine species stay extinct forever and cannot become part of any new technologies. This reason alone should explain why ocean protections are so important.

The World Wildlife Fund (WWF) has estimated that only 0.6 percent of the world's oceans receive any type of protection, yet oceans provide the Earth with vast biodiversity. Protecting larger regions of the ocean will give marine biologists a better chance to study many unique organisms that may play a role in future biotechnology. The WWF has explained, "The specialized adaptations of deep-sea organisms are not just interesting for interest's sake: an understanding of their biochemistry could also lead to biochemical, medical, and other advances." In this way, marine protections resemble land protections: They work best if people decide the region can provide some benefit to humanity.

METHODS FOR MONITORING MARINE LIFE

Marine scientists monitor marine life in three ways: (1) surface craft, (2) undersea manned craft, or (3) unmanned submersible equipment. Large research surface ships contain fully equipped laboratories for scientists to analyze fresh samples brought onboard. The Woods Hole Oceanographic Institution (WHOI) in Massachusetts uses four research ships, one of which, the R/V *Knorr,* carries four fully equipped laboratories. Both surface ships and manned undersea vessels contain sensors for measuring water conditions, and shipboard scientists collect data on coastal ecosystems and ecosystems of the open sea. The ship may also serve as a launch for undersea laboratories or undersea robots.

Undersea laboratories have become more sophisticated with the type of onboard analyses they can run, and some can hold up to four scientists to immediately study samples collected from the surrounding water. Undersea craft can carry scientists to the ocean floor and then employ specialized sampling equipment to bring collected specimens into the vessel. Undersea laboratories carry the following sampling equipment: (1) a grab sampler with a clamshell-type scoop for collecting floor samples; (2) a corer to drill into the ocean floor and collect sediment layers in a single column; and (3) zooplankton nets that settle on the ocean floor to collect tiny organisms, then retrieve the samples as the craft pulls them from the floor.

Deep-sea exploration relies on automated robots, undersea microscopes, and seafloor samplers. The following sidebar, "Undersea Robots," explains one of the most useful of these techniques for studying very deep habitats. The main advantage of these unmanned vessels lies in the fact that they can stay at very deep places for longer periods than would be safe for humans. WHOI and similar ocean research organizations use the following deep-sea study equipment:

- deep-diving vehicles that relay information from sensors to scientists on a surface ship
- remotely controlled vehicles for collecting data and samples from the ocean floor at depths to 14,764 feet (4,500 m)
- tethered vehicles that connect to a surface ship by cable and transmit images and sensor readings to the ship

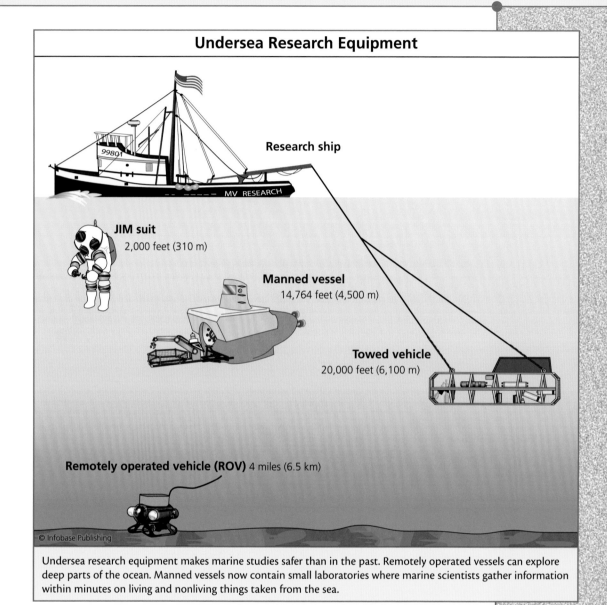

Undersea Research Equipment

Research ship

JIM suit
2,000 feet (310 m)

Manned vessel
14,764 feet (4,500 m)

Towed vehicle
20,000 feet (6,100 m)

Remotely operated vehicle (ROV) 4 miles (6.5 km)

© Infobase Publishing

Undersea research equipment makes marine studies safer than in the past. Remotely operated vessels can explore deep parts of the ocean. Manned vessels now contain small laboratories where marine scientists gather information within minutes on living and nonliving things taken from the sea.

From the samples brought out of the sea, marine biotechnology may be able to devise a number of innovations to benefit the environment. The inventions that will likely be the first breakthroughs derived from the sea are the following: (1) bioengineered microbes for pollution cleanup; (2) development of marine pharmaceuticals; (3) enzyme systems from ocean microbes; and (4) new materials.

UNDERSEA ROBOTS

arine biology includes certain challenges that biology studies on land do not confront. Not only do marine scientists find species that have never before been studied, but these scientists must also use specialized techniques for taking samples from the ocean while staying safe more than a mile below the surface. Manned underwater laboratories, called human-occupied vehicles (HOV), offer scientists a chance to do real-time studies on living and nonliving specimens at depths reaching 14,764 feet (4,500 m), or 2.8 miles (4.5 km). Scientists inside these small vessels can remain submerged for only about four hours, so that is why deeper explorations use unmanned equipment.

(continues)

The *Alvin* is one of the most famous undersea research vessels in use today. The vessel's titanium hull enables it to stay submerged under very intense pressures for up to 10 hours and at a maximum diving depth of 14,764 feet (4,500 m). The *Alvin* holds still and video cameras aided by powerful external lights. The vessel also uses two hydraulically powered arms for retrieving samples up to 200 pounds (91 kg); one arm is pictured here deploying bait traps to attract sharks. *(OAR/National Undersea Research Program)*

(continued)

In 2008 the University of Washington and WHOI announced a successful collaboration in launching *Sentry*, a robot that could operate at more than three miles (4.8 km) below the ocean surface. *Sentry*'s sleek contour with fins each holding a small propeller enables the vessel to swim like a fish or navigate like a helicopter. Unlike most other deep-sea robots, also called automated underwater vehicles (AUV), *Sentry* can break loose from a ship's remotely controlled commands and travel on its own through the water or to the ocean floor. The robot's power comes from 1,000 lithium batteries, its shell withstands extreme pressures, and the vessel can operate for up to 20 hours. The WHOI engineer Rod Catanach said, "*Sentry* is a true robot, functioning on its own in the deep water. The vehicle is completely on its own from the time it is unplugged on the deck and cut loose in the water."

Robots that remain under the control of shipside scientists are called remotely operated vehicles (ROV). These vessels act as a bridge between manned laboratories and independent vehicles like the *Sentry*. ROVs connect to the surface ship by a fiber-optic tether that can reach up to six miles (10 km) in length. Such vehicles contain video and still cameras, sonar equipment, and samplers equipped with lights. Scientists aboard ship operate all these instruments to study marine items at depths of up to 21,325 feet (6,500 m), or four miles (6.5 km). Undersea robots such as AUVs and ROVs operate for days at a time under great pressures in a single dive.

The complex nature of getting to deep-sea habitats and learning about their ecosystems partly explains why marine biotechnology may be the most difficult task in all of biotechnology.

PLANKTON

Plankton consists of diverse marine and freshwater organisms of a few micrometers in length to no more than an inch in length. Plankton belongs to two large groups: (1) phytoplankton, consisting of organisms that derive energy from photosynthesis, and (2) zooplankton, consisting of animal forms that must consume plants or other animals for nutrients

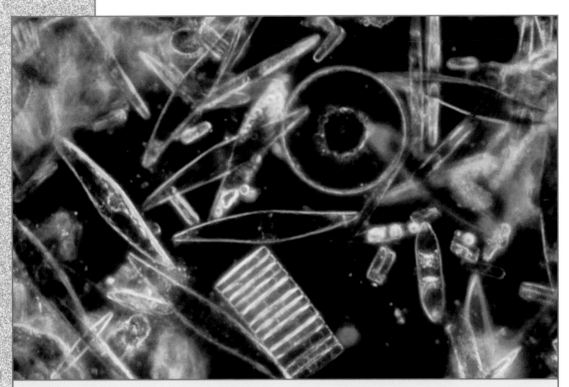

Plankton consists of a diverse collection of tiny organisms that act as the foundation for marine and freshwater food webs. Marine science and marine biotechnology pay attention to these organisms, their distribution, and ways to improve their vitality. The plankton shown here are diatoms; each species has a distinctive silica shell that protects the microbe. *(Gordon T. Taylor, Stony Brook University)*

and energy. Each type of plankton may number from several million to several billion organisms in a single quart of ocean water.

Phytoplankton represents one of the critical life-forms on Earth because these organisms act as producers in ecosystems. Producers are organisms that convert the Sun's energy to compounds that other organisms in a food chain feed upon for their nutrition. Earth ecosystems most familiar to biologists would not exist in their present form without phytoplankton living in fresh- and salt waters. Phytoplankton comprises algae, some fungi, and the single-celled microbes called *diatoms, dinoflagellates,* and *cyanobacteria,* which are described in the "Cyanobacteria" sidebar.

Zooplankton act as consumers in food chains; they cannot capture the Sun's energy by photosynthesis, so they must prey on other organisms for their nutrition. Zooplankton consists of protozoa, small crustaceans, worms, jellyfish, eggs and larvae of other aquatic animals, and krill. Krill encompasses all small marine crustaceans of about one inch in length that float near the ocean's surface and feed on phytoplankton. Krill serve as one of the most vital members of marine food chains and ecosystems.

Global warming and rising ocean temperatures potentially eliminate much of the ocean's krill and other plankton that exist best in a defined water temperature range. If these important foundations to marine ecosystems disappear, ocean biodiversity will also disappear in a short period of time. Though phytoplankton is essential for ocean ecosystems, some types of phytoplankton also create serious health problems when they grow to dense colonies in polluted offshore areas. These fast-growing, dense colonies called blooms harm marine ecosystems, kill or injure marine life, and produce toxins that can be lethal to people and marine mammals.

Marine biotechnology has not yet explored all of the potential uses of plankton, but these tiny creatures certainly offer a long list of topics for study, as, for example:

- Carbon absorption as CO_2 by photosynthetic phytoplankton annually produces 10^{15} to 10^{16} grams of carbon in a form that can be used by other organisms.
- Toxins that act on nervous systems may be developed into painkillers or other drugs.
- Oxygen-carrying capacity may be engineered for use as blood enhancers.
- Photosynthetic capacity may be harnessed as an energy source.
- Bioengineering may produce pollution-degrading plankton.

Because of the sheer numbers of plankton in the Earth's waters, these organisms may represent an untapped natural resource for the various types of biotechnology, perhaps even white biotechnology.

PLANT AQUACULTURE

Plant aquaculture is the cultivation of marine plants like crops rather than growing these plants unconfined in the wild. The common plants now part of aquaculture are the following: algae, seaweeds, kelp, sea lettuce, nori, and alaria. Several types of marine plants serve as foods, such as seaweed and nori, but these plants have also provided diverse raw materials for use in science and industry. Red algae provides an example; red algae produce agar that is used as a growth-medium ingredient in microbiology laboratories. Other algae produce the food stabilizer and thickener carrageenan, and other cultivated plants are raised as food for fish farms.

In environmental science, plant aquaculture could also serve as a resource for reestablishing marine environments that have been damaged by human activities. Aquaculture facilities containing plants from deep habitats may be designed someday to support marine life having value for biotechnology. Marine plants already have been used to make ingredients in cosmetics, drugs, and textiles, and they provide a good source of enzymes, minerals, and pigments. Depending on the attribute a particular industry seeks, aquaculture of bioengineered plants has a future as a primary source of raw materials.

The steps involved in bringing a marine plant from seedling to mature plant resemble the steps used in any nursery that houses terrestrial plants. *Aquaponics* is a term used for this blending of agricultural methods for raising new plants and aquaculture methods for rearing the plants in water. To begin, seeds already harvested from marine plants can be sown onto soil flats, which are shallow boxes containing clean soil. After seedlings sprout and grow into young plants, workers replant them to larger and larger flats or pots to supply more space and soil. As soon as the plants reach maturity, workers transfer them to small grow-out ponds and then larger grow-out ponds, where aquatic root systems develop.

Seaweed aquaculture already contributes to cleanup of polluted waters in a process called phytoremediation. These seaweeds absorb chemicals from the ocean to keep the chemicals from harming ecosystems. Aquaculture-grown seaweeds also offer promise for convenient production of marine-derived materials that are difficult to harvest from the ocean. In this scenario, the aquacultured seaweeds are easy to harvest and replenish.

Bioengineered aquatic plants other than seaweeds also possess unique abilities to produce materials for industry or products for consumers. Innovative aquaculture designs hold promise for use in flow-through drinking water treatment or wastewater treatment. Energy ponds represent another use for aquaculture-grown plants. Energy ponds are shallow ponds filled by actively growing plants for the sole purpose of capturing carbon dioxide (CO_2) and transforming it to energy in a form people can recover. These innovations remain in the exploratory stage, but they hint at the opportunities that people can derive from plant aquaculture.

ANIMAL AQUACULTURE

Marine animal aquaculture has received far more opposition than plant aquaculture, mainly because many scientists and the public doubt if it helps the environment. Animal aquaculture entails the raising of fish or shellfish in confined waters rather than sending boats out to harvest fishing grounds. There are two kinds of fish aquaculture: (1) fish farming, which raises fish explicitly for harvesting at an adult age, and (2) fish ranching, which holds certain species for part of their life cycle but releases them into the wild for breeding. For example, salmon lend themselves to fish ranching because they live part of their lives in marine waters and part of their lives (for spawning) in freshwater.

At least 20 percent of all fish sold comes from aquacultures; in some parts of the world, this percentage is much higher. In the United States, for instance, 40 percent of seafood is raised in captivity, and almost all of the following seafood comes from aquaculture: catfish, striped bass,

Atlantic salmon is native to the North Atlantic Ocean, but its population is at its lowest point in history. These fish have been severely harmed by overfishing, pollution, acid rain, habitat destruction, and dams that disrupt routes to spawning grounds. Atlantic salmon aquaculture has grown 35 times in size from two decades ago. Aquaculture has been credited with saving the dwindling salmon supply on the U.S. East Coast. *(William W. Hartley, U.S. Fish and Wildlife Service)*

trout, salmon, tuna, oysters, crawfish, clams, tilapia, shrimp, and mussels. Additional species that are increasingly being adapted to aquaculture are cod, haddock, and scallops. The biggest area of growth in aquaculture, however, is the farming of turtles, frogs, and sea urchins, shown in the following table.

The United States ranks about 10th in the world in the amount of fish raised by aquaculture. China, India, Vietnam, Thailand, Indonesia, Japan, Bangladesh, Chile, and Norway outproduce the U.S. aquaculture industry.

The WWF stated as early as 2003, "Aquaculture, or farming at sea of fish and shellfish, is set to overtake cattle ranching by the end of this decade. It is the world's fastest growing food sector . . ." Fish

Globally, aquaculture supplies about 50 percent of all seafood, and its proportion among all types of seafood production is increasing. In the United States, freshwater fish comprise more than one-third of all aquaculture; mollusks comprise about one-fifth. Between now and the next 20 years, the Woods Hole Oceanographic Institution estimates the global number of aquaculture fisheries will double from about 45 to almost 90. *(Caithness Community Web site, Scotland)*

WORLD AQUACULTURE PRODUCTION OF TURTLES, FROGS, AND SEA URCHINS

ANIMAL	1997 HARVEST, TONS (METRIC TONS)	2006 HARVEST, TONS (METRIC TONS)
turtles	52,521 (47,647)	232,879 (211,266)
frogs	1,994 (1,809)	95,831 (86,937)
sea urchins	82 (74)	95,460 (85,601)

Source: Food and Agriculture Organization of the United Nations (FAO)

aquaculture therefore helps provide income and food for poor regions of the world, but at the same time it may threaten the environment. Both cattle raising and aquaculture serve as sources of food protein, but many people criticize cattle ranching and fish farms for the same reasons, mainly because both enterprises produce large amounts of wastes that enter the environment.

The WWF is among several environmental organizations that oppose aquaculture because of environmental problems. The following list provides the main known disadvantages of aquaculture:

- pollution of surrounding waters from fish feed and feces
- pollution from antibiotics or chemical cleaners used in aquaculture
- vulnerable to pesticide pollution in runoff
- escape of nonnative species into the environment

CYANOBACTERIA

Cyanobacteria conduct photosynthesis, a rare feat among bacteria. Cyanobacteria represent a diverse variety of so-called blue-green bacteria because of their pigments—they used to be classified as blue-green algae. Cyanobacteria have been part of Earth's biological activities perhaps longer than any other living thing; their fossils date back 3.5 billion years. Cyanobacteria in marine environments are part of the phytoplankton upon which aqueous ecosystems depend, and like all photosynthetic organisms, cyanobacteria take in CO_2 and release oxygen. In so doing, they help reduce the buildup of greenhouse gases and global warming. Some cyanobacteria also absorb gaseous nitrogen and convert it to a form that other organisms can use when they ingest cyanobacteria. These bacteria therefore participate in three of the Earth's nutrient cycles: carbon, oxygen, and nitrogen.

Cyanobacteria possess certain capabilities that would seem to make them attractive for biotechnology. For instance, members of this group that have been exposed to heavy-metal pollution produce a protein called metalothionin, which binds the toxic metal and allows the microbe to safely store the metal in a sac inside its cell. In addition to pollution cleanup, cyanobacteria offer promise in the following areas: as a food protein source; as a source of enzymes, pigments, and vitamins; for production of compounds that kill disease-causing viruses, fungi, or other bacteria; or as a fertilizer due to the organism's ability to capture nitrogen gas in a process called *nitrogen fixation*.

- depletion of fish used as aquaculture fish feed
- inability of captured juveniles to renew natural ocean stocks
- concentration of parasites and disease
- predation by birds, sea mammals, and wild fish poses a potential threat to these species' health
- destruction of natural coastline ecosystems

Despite the troubles listed here, marine biotechnology has enabled aquaculture to offer at least two long-term benefits to the environment. First, aquaculture is an efficient way to raise fish as food, and bioengineered species further increase the yield of fish in smaller volumes of water than in wild conditions. Second, bioengineered species can be created to tolerate wide ranges in water temperature, resist parasites and diseases, and

In 2008, microbiologists at the University of Texas developed cyanobacteria to degrade cellulose and turn it into the sugars glucose and sucrose, which then serve as starting points in making bioethanol. These cyanobacteria power their reactions using the Sun's energy and, furthermore, require little nutrient input because they make their own nitrogen compounds. One of the project's researchers, David Nobles, pointed out, "The problem with cellulose harvested from plants is that it's difficult to break down because it's highly crystalline and mixed with lignins (for structure) and other compounds. The huge expense in making cellulosic ethanol and biofuels is in using enzymes and mechanical methods to break cellulose down. Using the cyanobacteria escapes these expensive processes." This application may turn out to be the most exciting of all of cyanobacteria's uses in environmental science.

Considering that Nobles and his collaborator R. Malcolm Brown estimated that the corn needed to produce ethanol for all U.S. transportation would need to cover 820,000 square miles (2.1 million km²), or almost the size of the entire Midwest, cyanobacteria energy certainly seems worth studying. Cyanobacteria, like all microbes, conduct their activities in an area vastly smaller than that needed by corn—microbes are microscopic organisms of only a few micrometers in diameter. These organisms will likely enter a new realm of energy-producing systems from biotechnology.

contain higher nutrient levels than wild fish. Aquaculture furthermore uses less fossil fuel than conventional fishing boats.

Marine biotechnology and aquaculture have many allies who feel this type of food production is vital and these industries can surmount their disadvantages. The National Aquaculture Association (NAA) has claimed, "Aquaculture's phenomenal growth and bright prospects can be attributed to an increasing demand for consistent, high-quality wholesome products by American consumers." The aquaculture industry understands that their methods have created problems for the environment and doubt in consumers. The NAA assured that "aquaculture practices of the past, both in the United States and in many places throughout the world, have changed and are continuing to improve. Production practices are most often determined by the availability of natural resources and various social and economic constraints. Aquaculture has been responsive to societal needs and continually develops and applies credible, scientifically sound information throughout the public and commercial domain." If these reassurances prove true, aquaculture will benefit society.

Tim O'Shea is owner of CleanFish, one of a handful of companies in the United States that touts its sustainable fishery methods. O'Shea gave the *San Francisco Chronicle* his blunt opinion in 2008 on the fate of the seafood industry and the future of aquaculture: "If sustainable seafood is to grow, you've got to be talking about aquaculture, because fisheries are not going to be expanding in the wild. And if we want positive change (in aquaculture), we need to recognize and reward the better practitioners." O'Shea has made it clear that aquaculture may become more a necessity than a luxury in the very near future.

Sustainable aquaculture probably sounds like an impossibility to fish farming's opponents. Consumers, environmental watchdog groups, and biotechnologists must wait and see if the aquaculture industry will provide clean and environmentally safe food for society without adding further damage to marine ecosystems.

CORAL REEF RESTORATION

Coral reefs are declining worldwide at an alarming rate. From 60 to 70 percent have been affected in some manner by human activities, making them some of nature's most endangered organisms, because of changing ocean temperatures, pollution, disease, and destructive methods of fishing the reefs.

Coral reefs contain hundreds of types of marine organisms that live together in a cooperative partnership, predominated by coral and dino-flagellates, a type of algae sometimes referred to as golden brown algae. Marine scientist John Reed of Florida's Harbor Branch Oceanographic Institution described reefs in 2005: "My first study, in 1976, was to see what lived in the coral, what used it for habitat. I began to study the invertebrates, and what I found out was that a small coral colony with a head the size of a basketball could hold up to more than 2,000 individual animals and hundreds of species, including worms, crabs, shrimp and fish. It was an incredibly biologically diverse environment that we had never known about before. By 1980, we realized that this [deep-water reef off Florida's coast] was a totally unique habitat found nowhere else in the continental United States. And possibly nowhere else in the world." Marine biologists continue to discover new facets of coral reef activity, but undoubtedly one of coral reefs' most important roles in ecology is as a source of biodiversity that contributes to overall marine health and activity. Coral reefs provide habitat for at least 25 percent of all marine species; the loss of coral reefs holds the potential to create serious damage to the world's ocean biome.

Reefs expand by building new coral on top of dead coral skeletons, but this growth is so slow—no more than three inches (7.5 cm) a year—that rapid environmental changes do not allow a reef to recover from damage. Unhealthy corals suffer coral bleaching, a change in the coral from its normal color to white. This bleaching slowly drains the life out of the coral.

Though a large part of blue biotechnology takes place for the purpose of finding new material that might be of use to humans, the biotechnology of coral reefs occurs for the purpose of saving coral species and not exploiting them. The National Oceanic and Atmospheric Administration (NOAA) has cited three areas where biotechnology can help conserve coral reefs. First, the isolation of genes that induce reproduction and growth may help restore declining coral reefs. Second, bioengineered marine bacteria can be put to work in cleaning pollution or protecting a reef from disease-causing organisms. Third, in methods already being used, engineered biosensors provide early warning of toxins that harm coral.

The Scripps Institution of Oceanography in California has gathered encouraging evidence that healthy coral reefs that have received some damage from climate change can recover to their original health. The Scripps professor Nancy Knowlton has said, "In a world of doom and gloom, it is important to know that reefs with exuberant coral growth and abundant

Coral reefs cover less than 1 percent of the ocean floor but support 25 percent of marine life. Coral reefs contribute to human and environmental health in the following ways: habitat and biodiversity; source of seafood for remote islands and larger commercial fishing; natural barrier against coastal damage from storms; and potential source of drugs and other beneficial materials. *(Mohammed Al Momany, NOAA/Department of Commerce)*

fish populations still exist." Coral reefs may best be served by a blend of new technologies and legal protections that keep human activities away from the world's remaining reefs. In lieu of high technology that has not yet come along, environmentalists have tried innovative ways to refurbish coral reef ecosystems. The following "Case Study: Florida's Tire Reef Experiment" recounts one such hopeful project for rebuilding Florida's coral reefs.

MARINE REMEDIATION SCIENCE

The marine environment presents as many obstacles in remediating its health as it does in studying deep habitats. This is because the ocean currents and tidal action can carry away materials that scientists purposely put into a habitat to aid its recovery. The oceans also receive an enormous but unknown amount of pollutants each day from runoff or from intentional dumping of wastes. Any remediation project must do battle with these challenges.

CASE STUDY: FLORIDA'S TIRE REEF EXPERIMENT

$\mathcal{S}$cientific answers to environmental problems take a long time, perhaps decades, to uncover. In the meanwhile, the public may become frustrated watching a habitat degrade under the weight of pollution, illegal harvesting, or other factors. In 1972 the people of Fort Lauderdale, Florida, attempted a plan to address two concerns about their environment: threatened coral reefs and waste buildup in the form of used tires.

Swept up in the good feelings from the first Earth Day only two years earlier, residents were happy to learn that the Army Corps of Engineers began dumping 2 million old tires in small bundles into the Atlantic Ocean. Ray McAllister, an ocean engineering professor at Florida Atlantic University, recalled, "The really good idea was to provide habitat for marine critters so we could double or triple marine life in the area." The tire reef took only a few days to create, and other places around the world soon followed Florida's lead with their own tire reefs.

Florida's well-meaning plan to restore destroyed reefs with an artificial reef made of old tires has not succeeded. Today, underwater salvagers are removing the reef plus the tires that have broken loose and drifted over a wide area. Divers have recovered as many as 2,500 tires in a single day, but many of the original 1 to 2 million tires remain underwater. Florida expects the removal project to continue at least until 2011. *(OAR/ National Undersea Research Program)*

The reef experiment's planners perhaps forgot an essential part of life in South Florida. It receives yearly and seasonal battering from hurricanes, tropical storms, and thunderstorms. Divers who monitored the tire reef found no life established there, but instead noted that little by

(continues)

(continued)

little the tires broke free and dispersed from the original reef site. In the decades since the reef-building, about 700,000 tires have drifted over more than 34 acres (0.14 km²) of the ocean floor. Tires have washed up on beaches; many have landed on healthy coral reefs and stunted their growth. Even worse, William Nuckols of the coast cleanup organization Coastal America said, "They [the tires]'re a constantly killing coral destruction machine." Unlike old ships that are sunk to provide reef habitat, tires do not provide a permanent site for life to take hold and grow. The tires possibly dispersed as each violent storm hit the Florida coast. Equally troubling, potentially toxic components may leach out of the rubber and poison any aquatic life that tries to settle on the artificial reef.

Divers now work to extricate tires half-submerged in the sand, tie them together, and have them hauled up to ships. The job will continue each summer until as many tires as possible can be freed and removed. Florida plans to haul the tires to plants that will either grind them up for road construction material or burn them for energy. Divers from the U.S. Army, Navy, and Coast Guard are doing most of the recovery, and county, state, and other agencies have helped keep the costs in check. Meanwhile, any tires floating free in the ocean threaten animals living in nearby natural coral reefs. More than three decades after the tire reef experiment began, McAllister admitted, "I look back now and see it was a bad idea."

Marine areas can be restored by making them legally protected areas that contain little or no human entry. Protections help habitats regain strength and also help marine animals rebound to normal population size. Other remediation methods resemble those used on land: mechanical removal of polluted sediments or oils or the use of remediation microbes and plants. These actions solve small, localized problems in the ocean but may not have much effect on the global ocean.

The overall health of ocean ecosystems at any place on Earth arises from the health of the plankton populations in the water, and this should perhaps be the central focus of any marine study. Phytoplankton absorbs CO_2 and by doing so helps control the accumulation of greenhouse gases and also provides a carbon and an energy source for ecosystems. As plankton die and decompose, their matter sinks toward the ocean floor and provides nutrients for the animal life in the deepest ocean realms. Plankton also need certain nutrients such as minerals for growth, and new technologies on the drawing board aim to help provide these nutrients.

Though the ocean provides most of the minerals plankton need, the mineral iron often occurs in low levels in ocean water. Plankton populations living in iron-poor water therefore cannot reach their full potential. Scientists have recently tried to adjust plankton growth by iron-seeding the ocean, adding iron compounds to several thousand square miles of ocean. The WHOI oceanographer Ken Buessler told *Time* magazine in 2008, "When we add iron, we create plankton blooms, but a lot of that just dies and decomposes [at the ocean surface]." The dead phytoplankton no longer absorb CO_2 from the atmosphere, but their decay and sinking into deeper water does serve the Earth as an enormous carbon store. This idea has yet to convince all scientists that an ocean can be manipulated to perform better, but if marine science does discover a way to restore ocean ecosystems, the job will likely be done by plankton.

Marine remediation science currently uses physical restoration methods, such as dredging, collecting oil spills, and designating sensitive areas such as MPAs. Biotechnology will likely play a bigger role in the future to help with cleanup of pollution using bioengineered microbes. Marine microbiology also contributes by developing microbes that clean up pollutants, signal the low-level presence of pollutants, or enhance the workings of nutrient cycles.

Marine remediation contains one big disadvantage and one promising advantage. The disadvantage comes from the difficulty of designing ways to affect an environment as massive as the ocean. The advantage arises from the incredible opportunity the ocean presents merely because so many different species live there that may be of use in marine biotechnology. In 2004, for example, the microbial geneticist Craig Venter led a team of scientists from Maryland's Institute for Biological Energy Alternatives (IBEA) in exploring the species of the Sargasso Sea, a unique ocean ecosystem near Bermuda in the Atlantic Ocean. They discovered 1,800 species of microbes alone, 150 new species of bacteria, and more than 1.2 million new genes. The geneticist Paul Falkowski of Rutgers University in New Jersey commented on Venter's study, "The total number of genes they found is mind-boggling." Surely some of these genes have applications for environmental projects in biotechnology.

Marine remediation science involves the restoration of healthy marshes, estuaries, and bays, such as the Chesapeake Bay pictured here. Estuarine, coastal, and shelf sciences represent specialties within marine science that focus on ecosystems in a region from wetlands fed by salty marine waters to the edge of the continental shelf. *(Estuarine Research Reserve Collection, NOAA/Department of Commerce)*

Venter discussed his marine studies in a 2007 Public Broadcasting Service (PBS) interview: "We found such incredible diversity, unexpected from almost any type of study. Every 200 miles [322 km], 85 percent of the organisms and [DNA] sequences were unique to that region." With the availability of millions of new genes, Venter suggested novel approaches that biotechnology has already begun since the time of that interview. "We're trying to create bioenergy by taking some of these genes and pathways and converting, for example, sunlight into hydrogen or sugar into a burnable fuel. We also . . . try and understand the chemistry of the oceans, where we capture back the carbon dioxide. We put 3.5 billion tons [3.2 billion metric tons] into the atmosphere. The ocean is the biggest carbon sink, more than a hundred billion tons [90.7 billion metric tons]. If we can shift that equilibrium slightly, we might be able to capture back more that's actually doing real harm to our planet." By shifting the total metabolism of the ocean—certainly no small task—scientists hope to find another approach to stop global warming.

CONCLUSION

Marine biotechnology is an emerging science that focuses on finding new substances from the ocean that can benefit humans, but it also seeks to devise ways to restore damaged ocean ecosystems. The world's oceans contain probably more biodiversity than any other place on the planet, and thousands (maybe millions) of marine species have not yet been studied or even discovered.

Science's lack of information on marine life results from the difficulty of studying the deepest parts of oceans. Oceanographic research vessels allow marine scientists to study samples aboard surface ships or in undersea vessels. A large portion of undersea studies, nevertheless, cannot be done by humans for extended periods of time. These studies must be assigned to robots that travel over the ocean floor taking pictures, detecting water conditions, and collecting samples. Scientists from aboard a surface ship control these undersea robots or some robots navigate on their own.

In marine biology, plankton serves as the backbone of almost all ocean food chains. Marine organisms depend either directly or indirectly on both plantlike phytoplankton and animal zooplankton. Plankton plays such a critical role in ocean health, any innovations in marine biotechnology will most likely relate to plankton activities.

Marine science also contains specialty areas that have all been developed with the objective of improving the ocean environment. Plant aquaculture, animal aquaculture, and coral reef restoration make up the three main specialty areas. But these disciplines each bring an assortment of unique problems that marine science must overcome if they are truly going to help the environment. Despite public concern about many aspects of biotechnology, marine biotechnology may well supply answers to the problems that aquaculture brings with it.

Coral reef and other habitat restoration may soon become an area helped by marine biotechnology. Many of the world's ocean reefs have been badly damaged by human activities. For this reason, marine biotechnology may be valuable in creating organisms that will clean reefs, protect them from infestation, or return native corals to health. Of all the different types of biotechnology, marine biotechnology may hold the most promise for finding new materials, novel enzymes, and unique living things. Marine biotechnology's future might also help defeat environmental challenges on land.

ALTERNATIVE MATERIALS AND PRODUCTS

Green building means the construction of homes, offices, or other structures in a way to minimize their ecological footprints. Also called sustainable building or ecological design, this field involves three areas of concentration: (1) building design, (2) conservation of resources, and (3) alternative materials. Green building design encompasses all features that encourage efficient use of energy, heat, light, water, and waste disposal.

Conservation of resources in green building reduces the total daily and yearly amount of water, energy, and materials a structure uses. For example, water conservation usually involves recycling used water, called gray water, as irrigation for gardens. Cisterns catch rainwater to conserve overall water usage, which also contributes to resource conservation. The third aspect of green building involves the materials that architects and construction companies use to create new buildings. The wise choice in materials for building and for consumer products can make a large and positive impact on the ecological footprint.

In addition to the ecological footprints described in the chapter 1, the use of materials can be thought of in terms of a plastic footprint, wood footprint, petroleum-product footprint, and so on. By selecting alternatives to nonrenewable resources, green building minimizes a building's total ecological footprint. Alternative materials come from three main processes: (1) recycling, (2) substitution, or (3) synthesis. Recycling common materials such as paper, plastic, wood, textile, and glass decreases the amount of new supplies that industries must produce to meet consumers' needs. Recycling then directly reduces the amount of natural resources harvested from the Earth and so indirectly lowers the ecological footprint by reducing the use of fossil fuels, greenhouse gas emissions, and energy waste. Substituting materials involves the replacement of rare or almost depleted natural resources with a material that is abundant. For example, bamboo now replaces many rarer woods, such as walnut and mahogany, for making furniture and flooring. The third technology, synthesis, usually takes advantage of biological processes—the reason it is called biosynthesis—that require less energy and produce less hazardous waste than conventional chemical synthesis. For example, hemp-based fabrics substitute for other textiles in upholstery to reduce the amount of bleaching, dyes, and chemicals that go into regular furniture.

Alternatives to traditional woods, plastics, and textiles have now grown into a major part of green design for buildings and also for consumer products. Several years ago, few shoppers received recyclable plastic bags, consumers had little opportunity to recycle product packaging, and old tires grew into monstrous mountains, often to be burned in the open. Today these materials flow back into production processes for a variety of goods and building supplies. This chapter explores the array of alternative building and design materials that have entered the market in the past few decades.

The chapter begins with a description of how materials can be reclaimed for entirely different uses in their new, recycled forms. It discusses the business and science of recycling and how it has grown from a cottage industry to a major part of sustainable enterprise. Different types of alternative materials are highlighted: woods, plastics, and consumer products. This discussion also includes technologies behind recycling. A case study in this chapter describes how alternative materials have been incorporated into a hotel in San Francisco, California. This chapter introduces biosynthesized

People have become accustomed to a variety of products made from recycled plastic. This deck and the furniture are made from plastic lumber. Plastic lumber comes in a variety of grades, colors, textures, and weights. (*Plastic Lumber Yard*)

polymers that are often the backbone of many alternative materials and that bypass the use of petroleum-based raw materials.

REUSE AND RECLAMATION

People with an interest in the environment use a process called the Three Rs to remind them of how to manage their ecological footprints. The term refers to the phrase "reduce, reuse, and recycle." Reduction involves selecting products that do not waste natural resources. For example, if a shopper were to purchase a cloth shopping bag rather than a paper grocery bag, that shopper reduces the use of natural resources—in this case trees. Of course, if the same cloth bag carries groceries week after week, it provides an example of reuse. Reduction and reclamation encompass all the techniques used by consumers or industry to get the most out of materials by putting them to new uses after the original use has been exhausted. The same shopper could use a paper grocery bag repeatedly for weeks and then use the worn bag to wrap a parcel to be mailed. This is an example of reuse.

Reuse of materials provides industry with an important cost savings, but as industries have become aware of their ecological footprints, reuse also helps in an ecological sense. The following table provides examples of how materials are reused today in construction.

Reclamation is the recovery of an item after it has been used and considered to be waste. Reclamation has slightly different applications from one industry to the next, however. Water reclamation and land reclama-

EXAMPLES OF MATERIAL REUSE	
MATERIAL	HOW IT IS REUSED
glass	countertop component, artwork
plastic	decking, outdoor furniture
rubber soles	doormats, carpet backing
tires	road-building component

tion involve the restoration of these resources. Water reclamation may include the collection of runoff filling a pond and cleaning it until it is safe to reuse, perhaps as drinking water. Land reclamation usually refers to the use of land that was once covered by water. Mine reclamation, by contrast, consists of methods to return an area polluted by mining activities to a safe environment for people and other biota.

In green technology, reclamation is a type of reuse in which a person seeks discarded materials and recovers them for a specific new purpose. As an example, very old, turn-of-the-century houses being remodeled receive visits from reclamation firms that salvage rare woods, marble, granite, and metal fixtures. This type of reclamation is sometimes called salvaging or, in building design, architectural reclamation. Usually reuse involves any activities to employ an item that would otherwise take effort to discard. Reclamation, by contrast, usually involves an active seeking of a specific material for a single purpose. The following table provides examples of reclamations.

Almost any item that a person views as having reuse value can be put to work again. Old houses from the early part of the 20th century have become treasure troves for used radiators, railings, fireplaces, garden

EXAMPLES OF MATERIAL RECLAMATION	
MATERIAL	**HOW IT IS REUSED**
wooden beams, staircases, flooring	new house construction, furniture, artwork, jewelry
metal fixtures, doorknobs, windows, doors, mantles	new interior design and construction
marble, granite	flooring, mantles, countertops, sinks
catalytic converters	platinum component used in making wire, alloys, jewelry
copper pipes, wires	brass and bronze alloys, gun metals
cars	parts

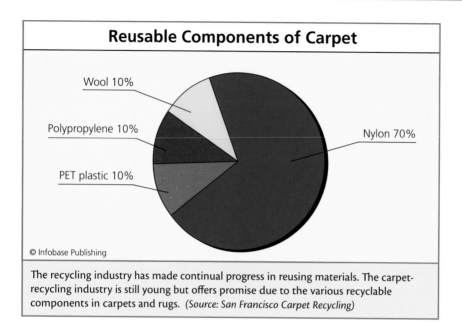

Reusable Components of Carpet

Wool 10%

Polypropylene 10%

Nylon 70%

PET plastic 10%

© Infobase Publishing

The recycling industry has made continual progress in reusing materials. The carpet-recycling industry is still young but offers promise due to the various recyclable components in carpets and rugs. *(Source: San Francisco Carpet Recycling)*

fixtures, timber, rock walks and fences, and stonework, plus smaller objects such as hinges, latches, knockers, and faucets. Architectural salvager Steve Drobinsky of Berkeley, California, told the television show *This Old House* in 2007, "This week one of the oldest mansions in San Francisco was being remodeled and they were removing marble sinks, cast-iron fire screens with stags and forests, a hand-carved walnut mantel—one leaf more than two feet [0.61 m] long, all hand carved! I mean, what could you get that could be better than that?" In addition to the fun of finding valuable old pieces for incorporating into a new home's design, this reclamation offers an excellent way to reduce the ecological footprint.

MUNICIPAL RECYCLING PROGRAMS

Municipal recycling programs have become the most efficient way of reusing items discarded as waste. Recycling began in much older societies when valuable resources could serve new purposes. Metals, pottery, fabrics, rope, and glass have been recovered from waste and recycled by artisans since the Middle Ages. Today's community-focused recycling is called municipal solid waste (MSW) recycling, and it progresses in two ways.

First, *primary,* or closed-loop, *recycling* occurs when a material is recycled to make more of the same material or product. Recycled paper that has been turned into new paper offers an example of primary recycling. Second, *secondary recycling,* or downcycling, occurs when recycled materials are turned into different materials or products. Plastic soda bottles reconfigured to make outdoor furniture illustrate secondary recycling.

MSW recycling follows a series of steps for collecting, sorting, and shipping wastes to places where they can be made into new products. In many communities, the sorting process begins at home, where families divide their paper, plastic, glass, and nonrecyclable wastes into separate containers. Waste haulers take the materials to centralized recovery facilities that further divide the materials and then consolidate each type of material into large loads to be sent to companies that use them as raw materials for new products. The more that MSW can be recycled in this way, the less waste must be incinerated or dumped into a landfill. Though incinerators and landfills have become much cleaner operations than they were before the 1980s, some of them nevertheless threaten the environment. Incinerators that produce hazardous emissions and leaky landfills present the biggest threats. Recycling helps minimize these potential problems.

The United States recycles about 33 percent of its solid wastes, which is about average among all countries. Switzerland and Austria recycle about half their solid wastes; Italy and Greece manage to recycle 10 to 15 percent of their wastes. Recycling technology has advanced to a point where industrialized countries have little excuse for avoiding MSW recycling. Recycling technologies now exist for the materials in the following list, and of these materials, iron, steel, and aluminum are the world's most recycled items. Electronics and appliances tend to enter another route in which their components are reclaimed to make new electrical items.

- iron and steel
- other metals (mostly aluminum)
- plastic
- paper
- glass
- motor oil

- batteries
- concrete
- fabrics
- wood

Certain difficulties make recycling harder than it should be if it is not managed properly. The main obstacle to efficient recycling relates to mixed materials, that is, metals mixed with plastics, contaminants mixed into aluminum, or different grades of plastics or metals mixed together. Industries need raw materials of specific grades in order to produce items of a desired purity. Each extra step that manufacturers must take to recover a material from a mixture costs money. Manufacturing costs rise, and consumers pay the price. The following table summarizes the advantages and disadvantages of recycling.

Recycling always makes sense for materials that are inexpensive to recycle, such as aluminum, steel, or paper, but may not be as sensible for difficult-to-recycle materials, such as computers. As recycling technology

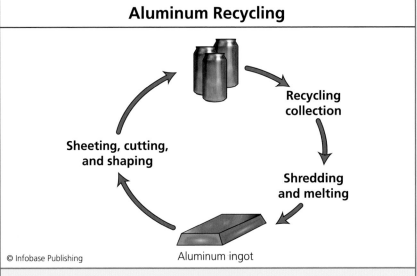

Aluminum Recycling

Recycling collection

Shredding and melting

Aluminum ingot

Sheeting, cutting, and shaping

© Infobase Publishing

Beverage makers began using aluminum cans in 1965, and today aluminum is the most efficiently recycled material. An aluminum can is 100 percent recyclable and can be recycled indefinitely. Within 60 days, a used aluminum beverage can is processed, made into a new can, filled with beverage, and returned to store shelves.

ADVANTAGES AND DISADVANTAGES OF MSW RECYCLING	
ADVANTAGES	**DISADVANTAGES**
• reduces ecological footprint in most cases • conserves natural resources • reduces greenhouse gas emissions • reduces wastes that must be disposed in landfills or incinerated • creates a new industry and jobs • creates awareness of sustainable practices • saves energy	• not cost effective for some materials, such as plastics and glass • reduces landfill and incineration-plant profits • uses fuel for hauling materials to sorters, from sorters to recyclers, and from recyclers to manufacturers • relies on separation at the source, households and businesses • creates complacency toward other sustainable practices • some processes create hazardous wastes; example: de-inking newspapers • requires extra work and time for collecting, hauling, sorting, and reprocessing

improves, recycling should continue to grow and contribute to reducing the ecological footprint.

SELECTING ALTERNATIVE RESOURCES

An alternative resource consists of any material that minimizes natural resource depletion. Alternative materials may be synthesized substances or biological substances. Two types of biologically made materials exist: (1) substances from newly discovered sources or (2) materials that substitute for a natural resource. Hemp plants produce the raw material for making fabrics that replace polyester fabrics and are abundant, so hemp plants can be considered a new source of an alternative material. Fast-growing woods provide an example of substitution by taking the place of endangered woods. A secondary benefit of alternative materials exists in the potential reduction of the carbon footprint. The production of alternative materials often creates a smaller carbon footprint than the production of conventional materials.

The construction industry was one of the industries to lead the way in using recycled materials on a large scale. Other companies now produce alternatives in clothing, personal care products, cleaning products, and foods. The following "Case Study: Orchard Garden Hotel" details how one business made a commitment to alternative choices. The following table presents numerous ways in which alternative materials occupy a role in construction and other uses.

Timber-certification programs include uses for both new woods that spare old-growth forests and recycled wood products. Timber certification involves a written statement from an agency that assures that wood products come from sustainably managed forests and that all the processes from logging to woodworking follow good practices in conserving energy, fuel, and natural resources. A certified sustainable wood product allows consumers to make a choice between responsible forest management and destructive styles of logging. In 1993 the U.S. Environmental Protection Agency (EPA) stated, "Several surveys indicate that a majority of Americans consider themselves to be environmentalists and would prefer to buy products with a lessened environmental impact when quality and cost are comparable." Consumers' desire for alternative materials took quite a while longer than the EPA's statement suggested, but alternative materials such as the ones listed in the preceding table have finally moved into the mainstream.

NEW WOODS

People have depended on wood as the main component of homes or furniture since antiquity. When the world population was much lower than it is today, timber could almost replenish itself after being harvested. In the 1900s, however, as the population began to grow at an ever-faster pace, trees began disappearing as a nonrenewable rather than renewable resource.

Deforestation worldwide has slowed from its torrid pace from the early 1900s to as late as 2000. Forests nonetheless continue to disappear for lumber, for the pulp and paper industries, and for burning for heat.

CHOICES FOR ALTERNATIVE RESOURCES	
ITEM	ALTERNATIVE CHOICE TO CONVENTIONAL PRODUCTS
cleaning	
disinfectants	vinegar, baking soda, lemon juice
general cleaners	vinegar/water mixture, baking soda/water mixture, soap and water, natural oils and waxes for furniture
home	
bedding	organic fibers, natural fibers, cotton/wool/latex mixes
construction	reclaimed, recycled, certified sustainable wood; granite or stone for walls and bases
countertops	slate, sealed concrete, recycled glass composites, wood fiber composites
flooring	slate, stone, bamboo, cork, certified sustainable wood, engineered or composite woods, nontoxic linoleum
furniture	bamboo, wheatboard, certified sustainable wood, reclaimed wood, recycled materials: cardboard fibers, newspaper, plastic, rubber
insulation	straw bales, recycled denim/cotton or polyester/wool, feathers
roofing	slate, galvanized metal, corrugated metal
outdoors	
driveways	gravel, crushed shells, recycled plastic lattice, natural earth
pesticides	biopesticides, natural predators, mixed cultivation
walkways	gravel, recycled rubber composites, recycled concrete

Africa and South America suffer the greatest rates of forest loss; some parts of Africa lose more than 10 percent in a five-year span. Deforestation also results in the release of carbon dioxide (CO_2) into the atmosphere, soil erosion, and loss of habitat and ecosystems. New materials that can take the place of old and undisturbed forests therefore play a vital role in

CASE STUDY: ORCHARD GARDEN HOTEL

The Orchard Garden Hotel opened in the heart of San Francisco, California, in 2006. The Orchard Garden attempted a new type of service for guests in a tourist city already full of hotels: a green hotel experience.

The Orchard Garden uses only chemical-free, citrus-based cleaning products; recycled paper; soy-based inks; soap and water to replace detergents; and energy-saving innovations, one of which is an energy cardkey for each room. The hotel's public relations office explains some of the building's features: "Guests activate the room's lighting and mechanical systems by inserting the card into a box near the door. When they leave the room, they remove the card from the box, turning off all systems except for one outlet, which can be used to charge laptops or cell phones." The hotel estimated that the cardkey system would save enough in energy costs to pay for itself in two years. Other hotels in Europe and Asia and even new sustainable homes have begun to include this technology for saving energy. These "smart" systems run air-conditioning, heating, and lights so that they automatically save energy whenever an electrical item is not in use.

In 2007 the Orchard Garden Hotel earned a new construction certification from the Leadership in Energy and Environmental Design (LEED) Green Building Rating System. This rating system encourages new construction projects to make ecologically sound decisions in construction materials, interiors, energy use, and consumable products (shampoo, soap, cleaners, etc.) used by the hotel staff and guests. The hotel's unique construction plan contributed to its certification. For example, 77 percent of the hotel's construction waste was diverted from landfills to other recycling and reclamation projects, and 22 percent of the building materials came from within 500 miles of the construction site, saving on fuel and emissions. The finished building includes a system for generating a small percentage of its own electricity. The hotel's general manager, Stefan Mühle, cited for the *San Francisco Chronicle* EPA research showing businesses can save about 50 cents per square foot by cutting energy use 30 percent.

Green, organic, or natural choices at home or when traveling may have once conjured the image of scratchy fabrics and uncomfortable furniture. The general manager Stefan Mühle assured the public in a *Newsweek* article on green lodgings, "You can have all the amenities you're accustomed

conserving biodiversity. Sometimes, timber companies can conserve old-growth forests by using substitute fast-growing woods that do not support the same complex ecosystems found in deep forests.

Wood technology has been making progress in supplying alternatives for homebuilders and furniture makers. Increasing choices have come on

to and still be green." He added for ABC News, "The fabrics throughout this entire room, whether it's the bedspread ... the curtains, the drapes, the sheets, the shower curtains, they all have recycled content in them. Anywhere between 10 and 50 percent." Mühle admitted that laws require chemical fire retardants to be part of the fabrics' composition. As for the furniture, Mühle explained, "All the furniture in the guest rooms, whether it's the chairs, the desks, the bed board, nightstands and so forth, that is all forest stewardship maple. It comes from sustainable-growth forests, so this is not from virgin forests." By selecting furniture made of sustainably grown woods, the Garden Orchard Hotel helps conserve undisturbed forests that have never been harvested. The hotel and others like it also show that alternative materials and prudent use of energy can create a structure that is both comfortable and ecologically sound.

Green hotels, such as the Orchard Garden Hotel with a room shown here, have been a growing sector of the hotel industry. The Green Hotels Association began in 1993 and today has several hundred member hotels. Each uses recyclable materials, water conservation methods, and biodegradable cleaning products. *(Jamie Colby)*

the market to conserve the three types of wood used in traditional construction described here:

- hardwoods—deciduous trees with broad leaves: oak, maple, cherry, walnut, beech, birch, cypress, hickory, and elm
- softwoods—coniferous trees: pine, hemlock, spruce, and cedar
- plywood—sheets of wood glued together so that the grains of adjacent sheets are perpendicular to each other

Hardwoods work best for making furniture and flooring because they are strong and durable. Softwoods and plywoods predominate in frames for new houses. Hardwoods and softwoods each contain species that have become threatened by overharvesting. Some types of maple, walnut, oak, and cherry trees have become scarce on the U.S. East Coast, while the vast coniferous forests of North America and the northern parts of the Asian continent are falling into an increasing state of threat.

This bedroom set illustrates the versatility of the variety of grass known as bamboo, which has become common for flooring and furniture. Bamboo has many other lesser-known uses, such as a material in roofing, fencing, packaging, tools, musical instruments, clothing, and paper. Bamboo even serves as a food in parts of the world. *(Style in Green)*

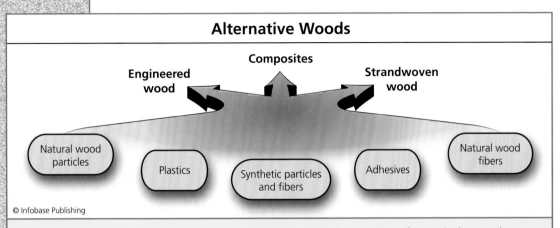

Alternative Woods

Composites

Engineered wood

Strandwoven wood

Natural wood particles

Plastics

Synthetic particles and fibers

Adhesives

Natural wood fibers

© Infobase Publishing

The science of making alternative woods has developed to give consumers a variety of strengths, hues, and textures. Additional new woods will result from advances in the recycling industry and in materials science.

New woods that have been devised to conserve the natural and threatened forests consist of engineered woods, composites, and strandwoven woods. Engineered woods, such as wheatboard, contain small pieces of woods from small-diameter, fast-growing trees that are usually grown on tree farms. Engineered-wood manufacturers press the small pieces together with glue in a heating step. The final product can substitute for plywoods or endangered solid woods. Green buildings use engineered woods that have been made with formaldehyde-free glues to avoid hazardous *off-gases* that these glues emit. Green builders prefer water-resistant, formaldehyde-free methylene-diisocyanate (MDI) glues. Composite woods have a structure similar to that of engineered woods, but composites combine waste wood fibers with recycled plastics. Composite woods can be painted or stained, and they do not rot, crack, or splinter. Strandwoven woods are a type of composite that contain wood fibers pressed into a dense material. Strandwoven woods differ from engineered woods in that strandwoven material contains long fibers lined up to form a natural-looking grain, while engineered woods have a mosaic appearance.

Natural woods also offer an alternative to old-growth timber. Bamboo and cork both grow quickly and have characteristics that work well in furniture and flooring. All woods are renewable resources; they all return after being cut down, but fast-growing woods offer *rapidly renewable wood* because these trees replenish their numbers in a short period of time. For instance, bamboo reaches a height of 60 feet (18 m) within months after being cut. Other woods take decades to reach maturity. The following table provides suggestions on woods that should be avoided in green building and good alternatives.

In the following table, "certified sustainable woods" refers to wood or wood products from sustainable forestry methods. These methods consist mainly of replanting cut trees, selective cutting that spares some mature trees, fuel-efficient methods of transport, and the use of nonhazardous materials.

ALTERNATIVE WOODS	
WOODS TO AVOID	GREEN ALTERNATIVE WOODS
temperate climate	
Alaskan cedar	certified sustainable woods
Douglas fir	cork
giant sequoia	engineered and composite woods
Sitka spruce	rattan
western hemlock	temperate-growth bamboo
western red cedar	wicker
tropical climate	
Brazilian pines	arariba (canary wood)
cocobolo	Brazilian angico
ebony	cancharana
ipê	chakte kok (Cuban or Jamaican mahogany)
mahogany	chechen
okoume	curupau
peroba rosa	katalox
ramin	t'zalam
rosewood	
teak, noncertified	

Bamboo Wood

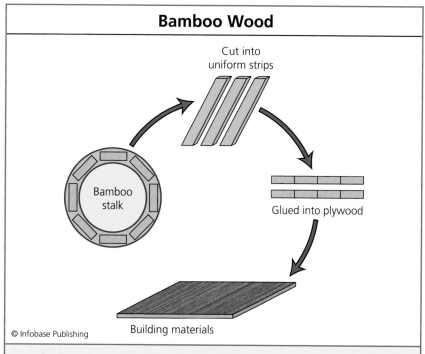

Cut into
uniform strips

Bamboo
stalk

Glued into plywood

© Infobase Publishing

Building materials

To make the popular alternative wood of bamboo, craftsmen cut long, flat strips from large stalks, trim the strips, and then bond them to make various types of lumber products. The world contains about 1,400 bamboo species, many of which are suitable for this process.

Bamboo is a fibrous grass that has become one of the most popular substitutes for rare woods. Bamboo ranges from one foot varieties to giant bamboo timber of more than 100 feet (30 m) and grows in a range of habitats, from jungles to mountainsides. Most commercial bamboo used in the United States and Europe comes from East Asia: the Philippines, China, Indonesia, Vietnam, and Japan.

Additional wood alternatives may be made of rattan or wicker, which are especially useful for furniture. Rattan is a fibrous material made from several species of palms and grown mainly in Indonesia. Rattan lends itself to furniture, but it does not hold enough strength for building construction. Wicker furniture contains different plant fibers, such as rattan and willow fibers, woven together into firm stalks. Additional choices for home and office furnishings are described in the "Ecological Furniture" sidebar.

CONSUMER PRODUCTS

Unlike 30 or even 10 years ago, green products exist as substitutions for hundreds of things people in industrialized nations use daily. Green choices extend away from home too in the form of biofuels; mass transportation; alternative means of travel, such as bicycling; and ecologically constructed office buildings. Green versions of consumer products exist for cars, appliances, electronics, home products, personal care products, and garden products in addition to the materials used in home construction. So many green choices have flooded the market for consumers, green has actually become overused, and perhaps misused.

Consumer products contain no single standard for declaring a product green or nongreen. The green product market includes such labels as "certified green" or "environmentally safe," but consumers may have a difficult time understanding the exact meaning of these terms. Different environmental organizations as well as product manufacturers complicate the picture because they have different meanings for the words "sustainable," "eco-friendly," "biodegradable," "recyclable," and so forth. Various environmental organizations use different criteria to determine whether they view a product as environmentally safe and, furthermore, use their own seals of approval for acceptable products.

"There should be a big caution to consumers," said Scot Case, director of TerraChoice Environmental Marketing, speaking to the *San Francisco Chronicle* in 2008 about various green designations. "Don't base your purchasing decision on some green dot unless you know what that green dot really means." The following table shows the major reputable organizations that rate products in regard to the product's ecological footprint. The footprint consists of two components: (1) the product's ingredients and (2) the manufacturing process used to make the product. In almost all instances shown in the table, the product label is intended to help buyers make the right decision when looking for green products.

Each type of product has its own set of criteria for reducing the ecological footprint. Even without knowing anything about various green ratings, most consumers understand that hybrid vehicles would be a better choice for the environment than gasoline-burning vehicles. The following list offers suggestions on the things consumers should consider when shopping for greener products. As this list shows, some products, such as electronics, offer few green alternatives. New green electronics

GREEN RATING ORGANIZATIONS		
ORGANIZATION	RATING SYSTEM BASED ON	EMPHASIS
Eco-label of the European Union	production and consumption characteristics of a product relative to its environmental impact	products, materials, and services used by consumers, retailers, and manufacturers
Environmental Choice program of Environment Canada	production, transport and distribution, and product/packaging ingredients	consumer and corporate products and services
Good Environmental Choice Australia	design and life cycle of product to minimize ecological footprint	the consumer product industry
Green Seal	U.S. production, transport and distribution, and product/packaging ingredients	cleaning products, paints and finishes, windows, doors, vehicle repair, and hotels
Japan Environment Association Eco Mark	design and life cycle of product to minimize ecological footprint	the consumer product industry
Taiwan Green Mark	product manufacturing, selling, promotion, and consumption	the consumer product industry
U.S. Green Building Council, Leadership in Energy and Environmental Design (LEED)	building construction materials and energy conservation	new construction, remodeling, and interiors for homes, schools, hospitals, and offices

may someday contain non-lithium batteries, circuit boards lacking toxic metals, and plastics protected with alternative flame retardants.

- cars—hybrid or electric rather than gasoline combustion
- appliances—energy-conservation ratings
- electronics—minimal packaging
- household products—minimal packaging, less chemical ingredients, recyclable plastic or no plastic

In a 2008 *Time* magazine piece on plastics, the reporter Bryan Walsh noted that the United States had produced 28 million tons (25 million metric tons) of plastic waste, of which 27 million tons (24 million metric tons) ended up in landfills. Walsh quoted the biologist Frederick vom Saal of the University of Missouri, referring to plastic-containing products: "These things are so ubiquitous that it is practically impossible to avoid coming in contact with them." Because plastics, even biodegradable forms, clog landfills and litter beaches, forests, and campgrounds and also emit chemicals that may harm ecosystems, many people have tried to reduce the amount of plastic they use daily. But as vom Saal suggested, it is nearly unimaginable to go through life without any products either made with or packaged in plastic.

Because of the difficulty of finding alternatives to plastic, environmental scientists have suggested that the best way to reduce the plastic footprint on Earth is to simply use less plastic. This can be done by reusing plastic bottles and containers, which has caused new worries to arise because of chemicals that leach out of the plastic with continued reuse, mainly

Stores such as Wal-Mart promote products that use sustainable packaging as well as environmentally friendly ingredients. Wal-Mart has a Packaging Scorecard, which gives consumers information on products that reduce the amount of paper, corrugated cardboard, and plastics they incorporate in their packaging. *(Enviromedia)*

bisphenol-A (BPA) for strengthening plastics, phthalates for softening plastics, and benzo(a)pyrene diol epoxide (BPDE) as a flame retardant. All three of these chemicals have been found in people's bodies using a technique called *biomonitoring* in which technicians analyze blood, hair, skin, and other tissues in a laboratory. Plastics that have long been assumed to be inert and safe may in fact be putting harmful chemicals into the body. BPDE, for example, interferes with the thyroid gland and the nervous system, and BPA and phthalates may act as *endocrine disruptors,* meaning they disrupt the normal function of the body's hormones.

Materials that serve as alternatives to threatened natural resources make up an important part of a sustainable lifestyle. In some cases, the choice to switch from a rare or declining resource to a plentiful resource is easy because the Earth provides many options. In other cases, few substitutes exist, and consumers are left with difficult choices for lowering their ecological footprints. Fortunately, new alternative products seem to appear regularly in industrialized countries where residents are most in need of reversing unsustainable habits.

PAINTS AND PLASTICS

Paints and plastics have for many years contained ingredients that cause serious health problems if inhaled or ingested. Paint's health hazards were discovered over many years and today most paint suppliers can advise consumers on the health advantages and disadvantages of each type of paint. The health hazards of plastics have only lately been exposed, and much remains unknown about the specific health effects of plastic components. Finding new formulas for paint and plastic presents a challenge to sustainable chemistry because they endure in the environment for a long period before decomposing; otherwise, consumers would not get the use they expect from these products. Persistence in the environment, however, is usually associated with chemicals that resist or harm the organisms called decomposers that help recycle the Earth's nutrients by decomposing organic matter. Chemists therefore must formulate paints and plastics that conserve natural resources and are safe for the environment yet provide long-lasting benefits for the people who buy them.

Indoor paints belong to one of the three following categories, from the most hazardous to the least hazardous: solvent-based paints; water-based paints, including acrylic or latex formulas and ecological or eco-paints. The relative hazards of different paints originate with components that emit *volatile organic compounds* (VOCs), which are carbon-containing compounds that evaporate into the air. Solvent-based paints produce the highest levels of VOCs, as high as 1.1 pounds (0.5 kg) per liter of air. These VOCs tend to release from the paint right after applying it to a surface and gradually decrease over time, but some paints continue emitting these compounds for several weeks to several months. Water-based paints produce less than half of the VOCs emitted from solvent-based formulas, and new eco-paints may emit as little as 0.03 pounds (0.015 kg) per liter of air.

Eco-friendly paints generally consist of about half water-based paint and half recycled paint, but specific formulas vary. Depending on the formula, an eco-friendly paint may emit VOCs in a range from 0.03 pounds (0.015 kg) to 0.4 pounds (0.2 kg) per liter of air. Eco-friendly paints also lack lead, which has been a common ingredient in paint formulas for many years with serious health consequences to people. Lead is a known toxic metal that causes permanent neurological damage in humans and animals.

A new generation of ecological paint contains no solvents and emits no VOCs. Linseed oil paint, for example, contains a mixture of natural linseed oil colored with either zinc oxide to make white or other metal oxides or natural pigments to produce colors. Ecologically friendly paints have fewer color choices at present than conventional paints, but sustainable chemists will likely increase the color choices available to consumers.

The plastics industry grew out of the 1862 invention of London chemist Alexander Parkes, who found that heating an extract from cellulose fibers allowed him to mold and shape the material, which then held the shape after it cooled. Parkes's substance molded better than rubber and also had a translucent appearance that other materials could not achieve. But plastic products did not enter the market in a big way until the plastics industry boomed in the 1950s and 1960s (*see* Appendix D). So important was plastic in replacing rubber and wood that plastic grew into an item used by almost every person in the industrialized world every single day. Over the next few decades, plastic replaced glass and metal in many products and began showing up as a component in cars, boats, furniture, decking, and carpeting.

Plastics, like paints, emit compounds that affect the environment and harm human health. Plastics differ from paints because plastics not only emit vapors but also fill up the environment with solid waste that decomposes slowly or not at all. To illustrate the persistence of plastic in the environment, a section of the Pacific Ocean about 500 nautical miles (926 km) from the California coast contains an estimated 100 million tons

Plastic Resin Ratings

 Polyethylene terephthalate (PET)
Water, soda bottles

 High-density polyethylene (HDPE)
Cleaning-product bottles

 Polyvinyl chloride (PVC)
Food trays, shampoo bottles

 Low-density polyethylene (LDPE)
Shopping bags

 Polypropylene (PP)
Microwaveable containers

 Polystyrene (PS)
Foam packaging for meat and poultry

 Other plastics
Cups, plates

© Infobase Publishing

Many people have learned that the plastic resin ratings on plastic containers indicate that the container is recyclable. Variations exist on the plastic resin ratings shown here. Recycling rating symbols also exist for glass, paper, corrugated cardboard, and aluminum. A University of Southern California student, Gary Anderson, designed the original triangle-like recycling symbol in 1970. The three arrows have come to represent the steps in recycling as well as the Three Rs: reduce, reuse, recycle.

(91 million metric tons) of partially degraded plastics. The floating waste dump has settled into an unusually calm area of the Pacific Ocean called the North Pacific Gyre. On a 1997 sailing trip from California to Hawaii, American oceanographer Charles Moore discovered what scientists now call the "Great Pacific Garbage Patch." The patch has now grown to the size of the entire area of the 48 contiguous United States.

In 2008 Moore described his first impressions of the massive garbage patch to Britain's *Independent*: "Every time I came on deck, there was trash floating by. How could we have fouled such a huge area? How could this go on for a week?" (The plastic wastes had perhaps gone undetected by planes and satellites for years because the mix is translucent and undetectable from above.) Chemists have since studied the patch and have learned that durable plastics and disposable plastics make up most of the partially digested and gooey mass. Chemist Tony Andrady at North Carolina's Research Triangle Institute said in the same *Independent* article, "Every little piece of plastic manufactured in the past fifty years that made it into the ocean is still out there somewhere." The Great Pacific Garbage Patch indicates that plastics are not the wonderful, all-purpose invention they were once thought to be. The following table summarizes the main plastics used today and their advantages or disadvantages to the environment.

Plastics have now become victims of their own success. Almost every environmental organization decries the overuse of plastic today because plastic components may lead to unhealthy ecosystems plus an obvious buildup of wastes. Recycled and discarded plastic items accumulate in landfills and on land and in the oceans. Plastic waste has injured and killed wildlife when it becomes stuck in the mouth or ingested. Wildlife and marine animals have starved to death because plastic objects have tied their mouths shut, caused injuries that prevent animals from finding food, or otherwise injured them.

Sustainable chemistry has three tasks ahead of it to correct the problems that plastic has brought along with its conveniences. First, chemists must find ways for giving recycled plastics new attributes so they can serve additional purposes. Second, polymer chemists must continue developing synthetic materials that hold up for a period while consumers use them and then degrade quickly in the environment. Third, biochemists seek biologically made polymers in nature that cause fewer health effects and yet give people benefits while not harming the environment.

PLASTICS		
PLASTIC	MAIN USES	ADVANTAGES (A) OR DISADVANTAGES (D)
high-density polyethylene (HDPE)	beverage bottles, grocery bags	A: recyclable and not known to leach chemicals
low-density polyethylene (LDPE)	bread bags, squeezable bottles	A: not known to leach chemicals and moderately recyclable
polycarbonate	baby bottles, microwave containers, eating utensils	D: leaches chemicals into liquids and foods
polyethylene terephthalate (PETE)	soda, ketchup, jelly, and peanut butter containers	A: recyclable and not known to leach chemicals
polypropylene (PP)	yogurt and margarine tubs	A: moderately recyclable D: hazardous during production
polystyrene (PS)	foam insulation, toys; expanded polystyrene (EPS) makes Styrofoam	D: contains carcinogens and difficult to recycle
polyvinyl chloride (PVC)	cling wrap and delicatessen food wrapping; water distribution pipes	D: leaches chemicals into food, contains carcinogens, and not recyclable

Source: David Johnston and Kim Master, *Green Remodeling: Changing the World One Room at a Time.*

NEW POLYMERS

A polymer is any large molecule of natural or synthetic origin composed of many repeating small molecules called monomers. For example, starch is a polymer made up of the sugar glucose monomers. The polymers polyethylene and rubber contain only one type of monomer. If more than one type of monomer goes into making the polymer, it may be called a

ECOLOGICAL FURNITURE

cological furniture is furniture that either contains alternative materials or has been made by sustainable manufacturing processes, or both. A shopper picking out new furniture considers the piece's woods, fabrics, leather, or plastic and other synthetic materials that went into it. Today consumers' main choice in ecological furniture resides in either alternative wood furniture or plastic furniture.

Formaldehyde is a key ingredient in the wood and upholstery of non-ecological furniture. Plywood glues containing formaldehyde hold wood components together in furniture frames; formaldehyde adds wrinkle resistance to upholstery fabrics. This organic compound emits a colorless vapor that irritates the eyes, nose, and lower respiratory tract, and the EPA has classified it as a carcinogen. Ecological furniture uses only woods and fabrics that do not contain formaldehyde.

Furniture made from recycled plastics serves two purposes in the environment: (1) it conserves real woods, and (2) it helps use up some of the plastic wastes building up in the environment. Manufacturers of plastic furniture usually collect plastics that can hold weight after they have been turned into furniture: milk jugs, soda bottles, water bottles, and food containers are all common ingredients in furniture. A typical formula used in making furniture con-

copolymer. Deoxyribonucleic acid (DNA) provides an example of a natural copolymer because it contains repeating units of a nitrogen base, a sugar, and a phosphorus-containing group.

Polymers also play a big part in giving plastics their familiar characteristics. Plastics are any substance other than rubber that possesses the following characteristics:

- made of many long polymers
- goes through a flowing or liquid state sometime during manufacture
- manufactured with heating or under pressure, or both
- can be molded into a desired final shape

The chemical characteristics of polymers determine the characteristics of the plastics made from them. The American Chemistry Council has summarized the general attributes of all plastic polymers, known as resins, as follows:

tains almost 95 percent recycled HDPE from recycled milk jugs and water bottles; fiberglass and coloring make up the remaining constituents. Manufacturers of plastic lumber make their product by mixing the components together while heating them, then squeezing the mixture through a machine called an extruder, which produces the desired shape. Extruded plastic boards can then be sawed to specific sizes in the same way lumber yards prepare wood products. Plastic lumber does not wear in harsh weather as fast as wood, and it requires no water sealants, stains, or paint.

Ecological furniture serves five purposes for aiding the environment. First, furniture made from alternatives to endangered woods helps conserve biodiversity and so protects habitats. Second, alternative materials such as wicker reduce the rate of deforestation. Third, furniture made from reclaimed woods helps reduce wood wastes that can make up almost 30 percent of the total tonnage of MSW. Fourth, lumber transport by trucks and cargo ships produces a large amount of the air pollution near port cities. Fifth, furniture that contains no formaldehyde or other materials that emit harmful vapors—adhesives, glues, stain-resistant treatments, dyes, and varnishes and lacquers—helps reduce the hazards of indoor air pollution.

1. resistant to chemicals
2. provide electrical and thermal insulation
3. lightweight and strong
4. can be made in a range of forms from sheer and flexible to bulky and rigid
5. offer a range of characteristics and colors
6. usually made of petroleum
7. used for items that have no alternatives from other materials

The last two items on this list provide two reasons why plastics create a problem in the environment. Traditional plastic polymers come from monomers present in crude oil, coal, or natural gas, but other formulas not made from fossil fuels are replacing them to serve two purposes: (1) to decrease the use of nonrenewable resources and (2) to create a more degradable substance. The sidebar "Microwave Chemistry" explains an innovation that may make a surprising end product out of used plastics.

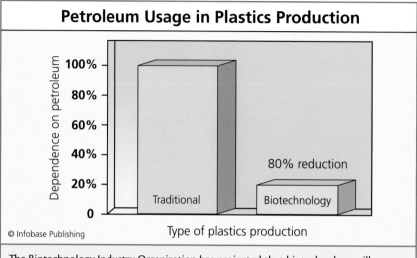

Petroleum Usage in Plastics Production

© Infobase Publishing

The Biotechnology Industry Organization has projected that biotechnology will lower the overall energy demand of the following industries: pulp and paper, textiles, wastewater treatment, and agriculture. Biotechnology might give the biggest boost to the plastics industry by greatly reducing the demand for petroleum, which is a raw material for current plastics. (*Source: Biotechnology Industry Organization*)

Two compounds have emerged recently as alternative plastics: polylactic acid from corn and cellulose from cotton. Cellulose acetate has been used as a natural polymer for making strong and durable plastics (screwdriver handles) as well as flexible items (package ribbon). Companies developing new polymers will be required to expand this rather short list of natural, nonfossil fuel polymers that deliver all the characteristics of today's popular plastics. Many medical devices—intravenous fluid bags, tubing, catheters, and blood collection items—have never been made with anything other than plastics, usually PVC.

Chemistry laboratories today turn out new synthetic polymers to deliver the following advantages: to replace chemical fire retardants; to produce lightweight concretes; to compose materials in nanotechnology; and to provide nontoxic surface finishes. Chemists can additionally formulate synthetic plastics to have the exact characteristics that work best in the final product. The major disadvantage of older synthetic plastic polymers resides in the fact that they do not decompose in nature. The next generation of synthetic polymers will therefore likely be molecules made from natural materials that decompose but also deliver the benefits of traditional plastics.

Chemists have already made progress using compounds in nature for new plastics. These so-called natural *biopolymers* may be altered physically or chemically to create a new *bioplastic*. Biopolymers for use as durable materials have been designed from natural starches, cellulose, plant fibers, tree fibers, feathers, and gelatins, plus a broad category of natural polymers called *polysaccharides*. Polysaccharides (also called complex carbohydrates) made from a single repeating monomer are called homopolysaccharides (starch, cellulose), and polysaccharides containing a mixture of monomers are heteropolysaccharides. In either case, all polysaccharides contain at least one monomer that is a sugar, such as glucose. The following table lists the main polysaccharides used in biopolymer production.

POLYSACCHARIDES		
POLYSACCHARIDE	**MONOMERS**	**SOURCE**
amylopection	glucose	starch
amylose	glucose	starch
cellulose	glucose	plant stalks
dextran	sucrose	synthetic; bacteria
lignocellulose	mixture of cellulose, hemicellulose, and lignin fibers	stalks of fibrous or woody plants
pectin	galacturonic acid, rhamnose, xylose, arabinose, galactose	plants
starch	glucose	corn, potatoes
xanthan	glucose, mannose, glucuronic acid, acetic acid, pyruvic acid	bacteria

Of the polysaccharides shown in the preceding table, all of the monomers are sugars except for glucuronic, pyruvic, and acetic acids. The monopolysaccharides composed of glucose differ by the ways in which the glucose units link together and if any branching occurs along the monomer chain. Xanthan is a heteropolysaccharide made up of five different monomers, so it can also be referred to as a pentasaccharide. Lignocellulose is a unique structure in nature that provides plant life with strength and consists of three components: (1) cellulose, (2) a diverse group of heteropolysaccharides called hemicellulose, and (3) lignin. Lignin is a large, complex molecule that gives woody plants and trees their strength. A small number of bacteria and fungi can slowly degrade this material in fallen trees, but all other organisms do not digest lignin.

The polysaccharide-type polymers discussed here hold more promise than other biopolymers for replacing plastics. (Proteins, DNA, and ribonucleic acid [RNA] are also biopolymers.) The polysaccharides can be altered to be almost indestructible, yet they decompose into harmless compounds. The chemical industry may invent new versions of natural polymers by adding branches or other chemical groups that change the properties of the final bioplastic. Bioplastics that degrade easily fit into a new category of sustainable processes called *biocompostable* materials. Biocompostable products consist of biologically produced plastics or paper products that have been formulated for a specific job but also have a formula meant to break down quickly in the environment.

Beginning in 2007, Berkeley and San Francisco, California, banned the use of polystyrene (Styrofoam), and other cities in the state and across the nation planned to follow suit. California entrepreneurs Allen King and Steven Levine stepped into the new market for biocompostable plastics by making plates, cups, bowls, straws, utensils, and other dinnerware from potatoes—they call their product Spudware. Plant residues left over after harvesting comprise a material called *bagasse* that has become a main raw material for biocompostable products. Other companies have now joined Spudware in making the best use of bagasse.

Allen King rightly predicted in a 2007 *San Francisco Chronicle* article that polystyrene bans in several cities would start "a huge surge of demand for alternatives from businesses that were using foam take-out containers." His business partner Levine added, "[Bagasse] replaces Styrofoam. You can microwave it, or heat it in the oven." Additionally, these biocompostable products are considered nontoxic, so there is no risk of accidental poisoning. Future biocompostable products will need to be faster at degrading than the current bagasse materials, but their advantages certainly endorse a closer look.

MICROWAVE CHEMISTRY

Microwaves consist of electromagnetic waves of the light spectrum with wavelengths outside the light spectrum that is visible to humans. The electrical and magnetic fields that make up microwaves possess two characteristics: (1) They have frequencies higher than radio waves and wavelengths greater than infrared radiation, and (2) they are oscillating waves, meaning they vibrate. Microwave vibrations change repeatedly from mostly electrical to mostly magnetic.

The oscillating microwave vibrations influence matter that comes into a microwave field. When a molecule enters a field, it becomes either electrically or magnetically polarized, meaning forces such as electric charge separate within the molecule. Polarized molecules try to vibrate in phase with the vibrating microwaves but cannot keep up, so the molecules begin jostling in a random fashion. As a result, the disorganized movement generates heat. Microwave ovens use this movement of molecules to heat food.

Chemists take advantage of the interaction with molecules and microwaves to hasten reactions in the laboratory and bypass less convenient heating methods. A firm in New Jersey has employed microwave frequencies to polarize plastic in this way and turn it into oil and combustible gas. Jerry Meddick, director of business development at Global Resource Corporation, explained in 2007 to *New Scientist* magazine, "Anything that has a hydrocarbon base [including plastic] will be affected by our process. We release those hydrocarbon molecules from the material and it then becomes gas and oil." Microwave technology therefore may have a future as a synthetic route to create new materials. If chemists successfully produce oil from plastics using microwaves, they will have invented a valuable tool for conserving natural resources and reducing waste.

COMPOST AND ITS USES

Compost is degraded nonanimal organic matter containing nutrients that can be used by plants. "Compost" conjures the image of decomposed vegetables and fruits that serve as fertilizer on farms or in gardens, yet compost also represents one of the easiest and oldest ways of creating a new material from decomposed matter. Such natural compost has now been joined by a new branch of chemistry that makes plastics called biocompostable plastics, meaning they are intended to break down like compost at the end of their life cycle.

Polylactic acid (PLA) and bagasse have become two leading choices for biocompostable plastic bowls, cups, and delicatessen containers. The *Smithsonian* magazine writer Elizabeth Royte, referring to PLA as "corn plastic" because it is derived from corn, explained, "Corn plastic is clearly easier on the environment [than regular plastic]. Producing PLA uses 65 percent less energy than producing conventional plastics ... It also generates 68 percent fewer greenhouse gases and contains no toxins." PLA represents one of the several new polymers expected to soon play a bigger role in sustainable chemistry.

Bagasse is a fibrous residue left over after juice has been extracted from sugarcane. It has been used mainly as a replacement for paper in plates, cups, and bowls and Styrofoam in take-out food containers. Like PLA, bagasse uses less energy and water to manufacture and produces less CO_2 emissions.

Developing sustainability will probably call upon many different technologies. New green buildings and products will require a combination of alternative materials such as new woods, synthetic plastics, and biocompostable plastics to conserve the world's natural resources.

CONCLUSION

Sustainability depends on the commitment of people and businesses to reuse, reduce, and recycle. These things may seem easy on the surface, but they will require a good deal of ingenuity from chemists to find energy-efficient ways to convert used materials into new products. Opportunities abound in this discipline, which can be referred to simply as alternative materials. Scientists endeavoring to create new alternative materials have concentrated on the following: woods, plastics, new polymers, and recyclable materials in MSW.

Alternative material use contributes to sustainability by conserving natural resources. Also, the production of alternative materials often emits less hazardous wastes than conventional materials. This is because, in many cases, alternative materials come from biological production processes rather than harsh chemical methods.

Alternative woods offer advantages to builders because they can be made stronger than natural woods. Of course, alternative woods also serve a primary role in reducing the harvest of old-growth forests, so they have a direct impact on protecting habitat and biodiversity and reversing global warming. Woods have a prominent future in building sustainability because they lend themselves to recycling, reuse, and reclamation.

Plastic recycling has helped reduce the amount of nondegradable materials in landfills and in the environment, but for the future, a better approach may be the replacement of plastics with new biodegradable materials. Plastic alternatives have already begun to be used in food containers, packaging, and some construction materials. Synthetic polymers, bagasse, and biopolymers will compose a larger amount of these items and other products in coming months.

Manufacturers that have produced plastics and paints have been major polluters for many years. Chemists working at these companies have already made encouraging inroads in three important areas: (1) the use of biological materials to decrease hazardous by-products; (2) development of biodegradable materials; and (3) paints, glues, and other substances that do not emit hazardous gases and metals.

Scientists who develop alternative materials have a busy future. Important needs include the following: faster-degrading biodegradable materials; a wider selection of biocompostable products; new woods that provide maximum strength and other desirable attributes; additional choices in safe paints, glues, and other chemical-based products; and development of microwave chemistry for plastic recycling.

Alternative materials and the products made from alternatives represent a vast opportunity in sustainable manufacturing. As additional materials come on the market, consumers may become overwhelmed by a growing assortment of green choices. Ecological labeling programs now exist worldwide to guide consumers toward products made in a sustainable fashion rather than products that remove natural resources from the environment.

For the near future, the alternative materials industry would be wise to target two objectives to help conserve natural resources: (1) increase the number of products it offers and (2) enhance the quality of alternative materials.

SUSTAINABLE COMMUNITIES

Sustainability refers to the ability of any system to survive for a given period of time. In environmental science, that system is Earth, or more accurately, it is the biosphere. People often overlook the second component of sustainability: lasting for a finite period of time. Sustainability cannot last forever, because all systems in the universe progress toward *entropy,* a state of disorder that contains insufficient energy needed to do work.

The principle of entropy supports the second law of thermodynamics, which states that energized conditions change spontaneously in the direction of a more stable, lower energy condition. The best way to understand entropy is to visualize a stone dropped into a still pond. The falling stone, its initial splash, and the series of waves it generates all carry relatively high energy. But the expanding waves and their energy dissipate as they travel out across the pond's water until the surface again becomes still. Eventually, all systems in the universe dissipate their energy in a similar manner.

A child playing with building blocks demonstrates how people deal with entropy. It takes a child much longer to construct a wall of building blocks than it does to knock the wall down. Before humans realized the world was not an infinite store of resources, they used up natural resources, discarded the leftovers, and wasted any excess, and humans have done this with alarming speed in the past century. In other words, people knocked down their building blocks. Even worse, humanity may not have the time and the remaining resources to replenish what it has depleted. The principle behind sustainability is to change the way people use the world's remaining resources for the purpose of conserving what is left of nonrenewable natural resources.

Without a more orderly and conservative use of natural resources, the Earth's state of disorder rapidly falls to a state that can no longer support life. Of course, people have learned to avoid this dramatic occurrence by conserving their resources. Unfortunately, as the world's human population pushes the limits of the planet's ability to sustain it, people must put extra effort into conservation.

Sustainable communities adopt several complementary activities that extend their future. These activities may be described in general as reducing, recycling, reusing, simplifying, sharing, and abstaining. All of these forms of conservation help people stretch their resources to last longer. Communities of any living organisms—a forest ecosystem, a city, a community of microbes living in a pond—sustain themselves by adapting to changes in their environment. In human communities, some examples of changes that loom or have already arrived are the following: depletion of the Earth's crude oil reserves; rising carbon dioxide (CO_2) levels in the atmosphere; and a reduction in urban living space.

Green technology's primary goal focuses on helping people use natural resources in a more sustainable fashion. The sum total of all these resources have sometimes been referred to as *natural capital,* because they represent the renewable and nonrenewable assets all living things have available to them to sustain their populations. Natural capital comprises the following nonrenewable resources: fossil fuels, nuclear power, minerals, and land. The renewable resources making up natural capital are the air, water, soil, biota, and energy from the Sun, wind, and tides. All the scientific specialties of environmental study seek to conserve natural capital in the same way individuals save their money—so that some will be available when they need it in the future.

Special regions of the world contain rich stores of natural capital compared with other regions, so it makes more sense for countries to work together in conserving resources rather than wasting those resources in one part of the world while people living in other regions are wanting. Sustainable communities described in this chapter refer to small units such as individual towns right up to the larger global community.

This chapter examines key aspects of sustainability by building it from the ground up or, in other words, from the local level. The chapter describes an ideal *eco-city,* which is an urban area that has maximized all the ways it can conserve resources. To date, no city has achieved 100 percent sustainability, but this chapter explores two places that have worked hard to reach that objective. The chapter also describes the economic and ecological indicators that help ecologists determine if the population is going in the correct direction toward greater sustainability. Finally, chapter 7 examines the future challenges to attaining a self-sustaining society.

THE ROAD TO SUSTAINABILITY

The United States awoke to the mounting damages being inflicted on the environment as early as the 1920s, when hunters and fishermen worried about a declining supply of game animals and fish. Presidents George Washington, Thomas Jefferson, and James Madison had spoken on the importance of taking care of the land that feeds people, but not until the westward expansion of white society did leaders begin to wonder if the United States could actually run out of resources. Thomas Cole wrote the following passage in *Lament of the Forest,* published in 1841, to convey his sense of loss.

> *A few short years!—these valleys, greenly clad,*
> *These slumbering mountains, resting in our arms,*
> *Shall naked glare beneath the scorching Sun,*
> *And all their wimpling rivulets be dry.*
> *No more the deer shall haunt these bosky glens,*
> *Nor the pert squirrel chatter near his store.*

Government leadership and strong activism would be required to turn the tide on rapid losses of natural resources as the 1900s unfolded. President Theodore Roosevelt devoted much of his energy to conservation and made it a priority for the country. Though people did not use the word *sustainability* as they do now, Roosevelt had notably used the correct terminology when he addressed Congress in 1901 on the need for forest conservation and land reclamation: "Wise forest protection does not mean the withdrawal of forest resources, whether wood, water, or grass, from contributing their full share to the welfare of the people, but, on the contrary, gives the assurance of larger or more certain supplies. The fundamental idea of forestry is the perpetuation of forests by use. Forest protection is not an end of itself. It is a means to increase and sustain the resources of our country and the industries which depend upon them." For

Sustainability comes from a wide variety of complementary actions. Recycling wastes, conserving fuel, choosing alternatives to nonrenewable resources, and reducing the use of personal vehicles are examples of the many choices available to people for building sustainability. *(Eric Langager)*

the first time in the United States, the federal government would develop an organized effort toward conservation.

Roosevelt's insightful opinions on conservation described conservation's two key objectives: (1) to manage a natural resource so that it is available for future generations but also (2) to manage conservation so that both nature and business can benefit. These are also the objectives of sustainable communities today. Any difficulties in building sustainable communities usually arise when the goals of environmentalists and the goals of industry do not align.

Sustainability happens only if environmental concerns take equal significance with industry concerns. Government must lead the way in balancing both issues. Today a number of international organizations guide governments in reaching a balance between environment and commerce. Two leading international organizations with this purpose are: (1) the Organization for Economic Cooperation and Development (OECD) and (2) the United Nations Environment Programme (UNEP). UNEP has stated, "The challenge of sustainable development . . . is to make trade, finance and globalization work for all members of society and the environment." Both organizations try to ensure countries build environmental protections into all new development projects, trade, and social programs.

Obviously, the real world imposes challenges on any plan to build the perfect sustainable community. But it may be useful nonetheless to design on paper a community that achieves perfect sustainability—an eco-city—in order to identify the achievable goals and the difficult goals.

BUILDING AN ECO-CITY

Eco-city design starts with the following eight factors that provide a blueprint for a sustainable community:

1. pollution prevention by cleaner production methods
2. waste prevention and reduction
3. habitat protection
4. restoration of damaged habitat
5. most efficient resource-use methods available
6. population stabilization
7. maximum conservation possible of renewable and nonrenewable natural capital
8. processes that allow businesses to earn a profit

All of the eight steps in building sustainability are achievable, but some require greater cultural changes than others. For example, societies can improve resource use and escape poverty by stabilizing their population growth rates. People do this by controlling the fertility rate, which is the number of births per woman. But will people agree to control the fertility rate? Family size often correlates with cultural traditions. Religious beliefs also influence decisions regarding family size and the use of birth control. Birth control offers the best opportunity for keeping population size within the Earth's carrying capacity.

World fertility rates have been falling during the past 20 years, which has been an encouraging sign for population control, but the Central Intelligence Agency (CIA)'s *World Factbook* reports that more than 30 countries continue to have fertility rates greater than five children per woman. The region of the world struggling with the highest fertility rates is in Africa: Mali, 7.34 children/woman; Niger, 7.29; Uganda, 6.81; and Somalia, 6.6. Afghanistan has the highest fertility rate of countries outside Africa, about 6.6 children per woman. The United Nations chairs programs for countries such as these for family planning while also supporting health care for pregnant women, obstetric care, and childhood health. Population stabilization in some parts of the world nevertheless remains an ongoing challenge.

Eco-city design must also include opportunities for industry to prosper. Business owners may need to make hard choices if certain sustainable operations cost more than traditional methods of doing business. For many years, manufacturing found it easier to discard wastes than to reuse or recycle them. In the 1960s, when recycling programs were nonexistent in many industries, companies no doubt found that ignoring wastes costs less than new technology to reuse waste materials. Sustainable communities must include sustainable manufacturing as part of their plans, but to make this work, a support industry that recycles and reuses wastes must be part of the plan. Elizabeth Lessner of the Central Ohio Restaurant

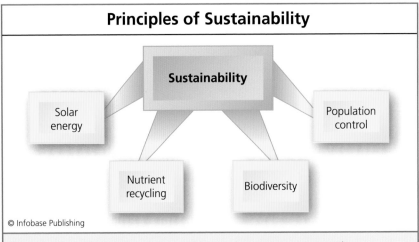

Principles of Sustainability

Sustainability

Solar energy

Population control

Nutrient recycling

Biodiversity

© Infobase Publishing

Some decisions that help in the sustainable use of resources are easy, such as community recycling programs. But long-term sustainability also requires large changes in the ways society uses energy, protects ecosystems and habitat, and controls population. Each of these issues must be addressed to support today's and the future's human population.

Association mentioned to the *Columbus Dispatch* in 2008, regarding the costs of recycling glass bottles, "Businesses generally want to do the right thing. It's just learning to do it and figuring out how to afford it." Waste managers will play a growing role in helping industry reduce wastes from the moment they begin operating.

An eco-city would do best by mimicking sustainability in nature. Plant and animal species sustain populations by following four interconnected principles: population control, nutrient recycling, optimal use of solar energy, and maintaining biodiversity. By regulating these four factors, wildlife and plants ensure that they use natural resources to the best advantage while leaving enough to sustain the next generation.

A planned eco-city should emphasize the characteristics described in the following table. Other underlying factors would also influence different eco-cities. For instance, a city having a warm, rainy climate could put less emphasis on heat conservation and water retention than a city in a drought-stricken area with cold winters.

Building eco-communities in impoverished areas represents one of society's biggest challenges. Impoverished areas tend to lack enough of their own resources to meet their economic needs. In many parts of the world, poverty has forced people to destroy habitat, deplete endangered natural resources, and give up basic sanitation and waste control. Poverty then leads to the decline of the environment's overall quality. Therefore, an eco-city must make provisions for reducing or eliminating poverty.

Ordinary citizens have little control over widespread poverty and national economics. Since the first Earth Day on April 22, 1970, however, environmentalists have stressed the importance of the individual for inspiring big changes. This is the philosophy behind grassroots groups—big changes take place only if they occur from the bottom up. If more and more people take small actions that help the environment, those actions coalesce into critical change for the environment. Almost every U.S. community now has a grassroots organization that works on environmental issues. Residents can find these groups through local waste management companies, local government offices, schools, or libraries.

The sidebars "Curitiba, Brazil" and "Tapiola, Finland" give examples of how cities try to build change by starting at the local government level. These changes do not always occur easily, and the road to an eco-city has its share of successes and shortcomings.

CHARACTERISTICS OF AN ECO-CITY	
FACTORS	APPROACHES
pollution prevention by cleaner production methods	reduced packaging; biodegradable packaging and products; low-emission/low-discharge processes
waste prevention and reduction	reduced packaging; take-back programs for unused product; reuse of manufacturing materials; reuse of wastes as raw material for another industry
habitat protection	protected wilderness; no-growth zones; maintenance of wildlife corridors; pollution and chemical-pesticide prevention
restoration of damaged habitat	bioremediation; land reclamation to natural state; pollution cleanup
most efficient resource-use methods available	no-tillage agriculture; sustainable forestry; alternative materials for fossil fuels and ore; waste-to-energy processes
population stabilization	family planning; birth control education; national population policies; reducing poverty; elevating status of women
maximum conservation possible of renewable and nonrenewable natural capital	renewable solar, wind, geothermal, or tidal energy; waste recycling; resource reuse and reclamation
processes that allow businesses to earn a profit	locally produced products; energy-waste exchange between industries; telecommuting; reduced packaging; emphasis on quality over quantity; ending short-term profit reports for investment community; emphasizing long-term profitability and eliminating quarterly profit estimates

PRESERVING GREEN SPACE

Urban green spaces such as botanic gardens, parks, and tree plantings provide more than a sense of nature in a metropolitan whirlwind. Green spaces provide cities with valuable environmental needs, such as the following:

- trees—shade for climate control and energy savings; removal of greenhouse gases; oxygen production; nutrient restoration to soil; water storage

- shrubs and other low growth—shelter and hiding places for small mammals and birds; attract pollinators; water retention and reduction in soil erosion

- wildflowers—draw insects and reptiles that keep invasive species in check

- grasses and woodlands—migration corridor for animals; plant and animal habitat; reduction in light and noise pollution

- varied vegetation—reduces population density, which may benefit health; fosters biodiversity; supports bacteria and fungi that degrade wastes

Australian authors Nicholas Low, Brendan Gleeson, Ray Green, and Darko Radović explained in their 2005 book, *The Green City*, "The task of integrating and encouraging nature to thrive in the city means that [you] have to look simultaneously at the big picture (the scale of the entire landscape) and the small picture (the scale of neighborhoods and individual developments)." Sustainable communities simply cannot survive in a healthy manner without green spaces, because green spaces conserve energy, store water and carbon, cycle nutrients, maintain plant and animal diversity, control wastes, and control pollution.

The nature in an urban setting provided by green spaces offers much more than shady places for people to walk and playing room for children and pets: Green spaces maintain the health of the air, land, and water in an urban center. Green spaces also play a less obvious role in human health. This is because they provide places where urban dwellers can experience a setting that removes them from the stress of the workplace. Kelly Quirke, a proponent of urban green spaces, told the *San Francisco Chronicle* in

2006, "As more of us live in cities and as we lose more open space, people are getting more and more of their experience in nature in their urban environment. This means the urban forest becomes ever more critical."

Even with the benefits, planting trees as new green space requires planning so that it does more good than harm. Ted Williams of the Audubon Society wrote in *Audubon* magazine in 2007, "The public doesn't understand that forests and trees are not the same thing." As Williams suggests, green spaces must contain a variety of plants, shrubs, and trees of diverse species and of different ages. This variety helps cycle carbon through the

CURITIBA, BRAZIL

The city of Curitiba in southern Brazil has throughout most of its history tried to keep up with population booms. European settlers used Curitiba as their port of entry to South America from the mid-1800s, and from the end of World War II into the 1960s, the city's population grew 10fold. The population spurt brought troubles involving traffic gridlock and social decay. In 1969 the city's government and business leaders decided to redesign how Curitiba would work. They decided to begin Curitiba's transformation by starting with its transportation system. Curitiba's leaders believed that improving transportation efficiency might well open opportunities to the city's residents and so raise overall living conditions.

Curitiba's plan included a coordinated train and bus system that reduced overlaps in routes. The transit plan also included what designers called a "trinary road design" consisting of three basic transportation modes. First, planners minimized city traffic by devoting a central artery to buses that made stops at points between the suburbs and the city center. Second, local car traffic stayed on wide, fast-moving, one-way streets. Third, fast-transit car lanes and bus express lanes radiated out from the city center to the inner districts, the outer districts, and the industrial district that ringed the city.

The trinary system became known as the Curitiba Master Plan, which sped the movement of shoppers, workers, and students to and from their main destinations. The master plan also made provision for undeveloped green areas and city parks plus citywide services such as recycling pickup. Curitiba recycles 70 percent of its paper and 60 percent of glass, metal, and plastic. Government and social services moved into decentralized local offices, called "citizenship malls," rather than a central downtown headquarters, a design that also alleviated traffic flow into the city center.

atmosphere at maximum efficiency. Such green space diversity also fosters habitat and animal diversity.

Green spaces require good choices, as mentioned, and sound planning. An entire eco-city presents a difficult challenge for city planners because eco-cities demand even more complex plans than small green spaces among city blocks. Eco-cities must balance people's needs with activities meant specifically to protect the environment. The sidebar "Case Study: Lessons from Biosphere 2" shows the troubles that arise in trying to re-create a miniature environment.

Curitiba's success came from a transportation system that addressed the area's poor. Special buses converted into traveling grocery stores and pharmacies go to the poorest sections of the city. The traveling market's produce includes extra crops from nearby farms that would otherwise be wasted. The transportation master plan enabled access for the poor to child nutrition centers and day care, education centers, small business loan offices, and health clinics. Schools put as much importance on teaching ecology as they do on other subjects.

Curitiba's three-time mayor Jaime Lerner has been credited with leading the transformation of a previously depressed area into one of the world's models in sustainable living. "I always felt a great connection with the street," he told the *Guardian* in 2008. "My dream was to be an architect. I saw things happening that I thought were wrong. They [previous city leaders] were destroying the city's history, opening up big roads that wiped out the whole memory of the city, planning the city just for cars. The [eco]-city of Curitiba became a reference for doing exactly the opposite of what other cities are doing. Other cities were building big bridges and freeways, and we were making pedestrian streets. Many cities were building metro systems, and we started our own transport system." Perhaps most amazing of all these accomplishments, Lerner's team made the changes so rapidly they outpaced the normally sluggish bureaucracy.

Curitiba's success came from forward-thinking leadership, proactive business and social groups, and residents willing to make the changes required of them. The residents accepted new ways to commute and learned to seek local services rather than services in the city center. Finally, Curitiba took the unique approach to solving some of poverty's problems by building a transit system that directly addressed the needs of the poor and low-wage workers. Curitiba demonstrates that any city's choice to become an eco-city has its own unique issues and solutions.

TAPIOLA, FINLAND

In the 1950s Finland and the rest of Scandinavia staggered under debt from World War II. Housing was scarce, and what was available barely met the needs of its residents. In densely populated Helsinki, large families wedged into a single room. A Finnish family agency happened to hire a young lawyer named Heikki von Hertzen to help work on solving Finland's housing problem. Von Hertzen had been charmed by the garden cities of England in which towns included trees, shrubs, and plants in all of their expansion plans. He made up his mind to do the same to revitalize Finland.

Von Hertzen managed to convince other agencies of his vision for the city of Tapiola. Tapiola would rebuild in a way that would include jobs for all social groups, transportation access to those jobs, ecologically planned housing, and plenty of greenery. The Helsinki news reporter Antti Manninen remarked in 2003, "Asuntosäätiö [the Housing Foundation of Finland] wanted to make Tapiola into a diverse dwelling area into which all comers had the chance to move, irrespective of their income or professional background. Apartment blocks, row and terrace houses, and detached properties all went up side by side with one another, and all in harmony with the unfelled forests, the meadows, and the planted gardens." Tapiola endeavored to rebuild itself as an eco-city from the ground up.

Von Hertzen's team of architects envisioned Tapiola as Europe's template for nature-conscious living. (In Finland, Tapio is the name for the ancient god of the forests, and Tapiola refers to the god's domain.) The project soon began winning building competitions: Aarne Ervi designed a downtown area that drew praise from European and U.S. architectural groups; the firm Pietilä, Raili and Reima designed a wooded residential neighborhood using concrete, wood, and even

ECOLOGICAL ACCOUNTING

Ecological accounting, or ecological economics, estimates the accumulation or depletion of resources for sustaining the world's biota. Simply, ecological accounting tells scientists whether people are living in a sustainable fashion.

Natural capital contains two components: natural resources and human-made resources. (Human-made resources differentiate further into human resources [people], built resources [construction], and cultural resources [social programs].) Both components of natural capital must be conserved in order to extend a population's sustainability into the distant future.

army fortifications left over from the war; and landscape architect Jussi Jännes created the city's parks and meadows. At least a dozen architects worked on Tapiola's rebirth. Since then, architects have added innovative office buildings and multistory residences containing local woods in combination with reclaimed concrete, metal, or glass.

Garden cities have since used Tapiola as a model for self-contained communities that decrease traffic; include meadows, gardens, and forestland; and balance industrial, residential, and agricultural areas. Of course, fast-growing urbanization takes its toll on cities, even Tapiola, which now has new buildings that fail to conform to von Hertzen's vision.

The wooded eco-city, or garden city, has been a strange idea to understand for many people who have known nothing but traditional city living their whole lives. Today in Finland, about one-fifth of the country's people live in city apartments. Finnish author Johanna Hankonen wrote in 1998's *Out of the Forest*, "The world-famous Tapiola Garden City became a pilot development for family policies . . . It was possible to build high apartment blocks in the forests as long as this allowed the residents access to the natural world." But Hankonen pointed out, "In 1967 the radical young generation branded Tapiola an unsuccessful 'forest city' and demanded that no more should be built." Some of the city dwellers never became accustomed to living away from city lights and noise for the quiet and solitude of the garden cities.

Tapiola shows that eco-cities present a multitude of challenges for city designers and for residents. Sustainable communities must overcome resistance from city dwellers who have never envisioned a city as anything but concrete and asphalt.

Ecological accounting resembles financial accounting. For example, a person might possess $50,000 inherited from an uncle (natural resources) plus a savings account that holds $15,000 (human-made resources). Without a job but with monthly expenses of $3,000 each month, this person could sustain her lifestyle for 21.7 months. By getting a job that pays $2,000 a month, this person's sustainability extends to 65 months, or almost five and a half years. A job paying $5,000 a month extends sustainability even more and may even accumulate an inheritance for the person's children. Ecological accounting works in the same way by conserving resources so that they are not drawn down so quickly that a population cannot sustain its current lifestyle.

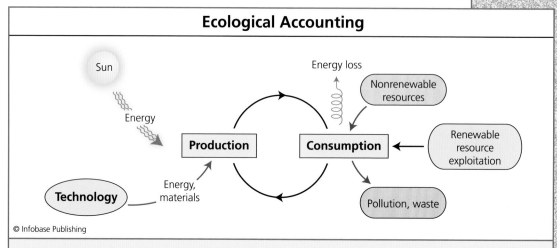

Ecological Accounting

© Infobase Publishing

Perhaps the best way to understand ecological accounting is by diagramming the routes taken by energy, resources, and wastes in human activities. Technology can contribute to new methods of production and new materials and so conserve natural resources. Technology also leads to new types of waste and pollution treatments.

The ecological footprint provides the most familiar calculation for estimating sustainability. By accounting for the energy, land, water, air, and materials needed to live and the actions needed to eliminate wastes, the ecological footprint tells a fairly accurate story on how well humans sustain their resources. Some parts of the world fail at sustaining resources. In order to make up this ecological deficit, a region must either take resources from other parts of the world (like borrowing from a rich aunt) or invent new technologies to create more resources (getting a higher-paying job).

What if no rich aunt or better job exists? In that case, countries select a third option, called overexploitation, which is described by the Australian author Ron Nielson in 2006's *The Little Green Handbook*. "If an ecological footprint of a country is larger than its ecological capacity, the country has an unsustainable style of living and lives on an ecological deficit—that is, beyond its means. However, this does not mean that the country is suffering deprivation. An ecological deficit can be offset by drawing on the resources of other countries or by overexploitation of domestic natural resources. The overexploitation of natural resources occurs everywhere, even in countries without an ecological deficit. It is an undesirable course of action that achieves a temporary solution to the problem of excessive consumption. In the long run it is economically devastating." Sustainability's primary goal is to rein in overexploitation.

Depletion of nonrenewable resources offers an obvious example of overexploitation. Though no one knows exactly how much crude oil the Earth still holds—some oil companies guard their estimates—many scientists think that humans have already drawn down more than 50 percent of certain reserves. No one seems to know for sure the status of the world's remaining oil reserves, but the oil industry may have already reached its peak production. The *Washington Post* writer Steven Mufson wrote in 2006, "Exploration and production in deep-water areas have become more important as production from older fields on or close to shore begins to decline." New oil fields have been harder to find and more expensive to tap since the late 1980s.

The oil fields in Alaska's Prudhoe Bay held North America's largest oil reserve when they were discovered in 1967, but as Sonia Shah wrote in her 2004 book, *Crude: The Story of Oil,* "By the late 1980s, the better part of Prudhoe Bay's oil had been drained . . . From a production peak in 1988, each well's output declined by 20 percent every year, despite a spate of drilling and the injection of gas and other chemicals to encourage more oil to ooze out." Oil may someday become the most critical of all overexploitations.

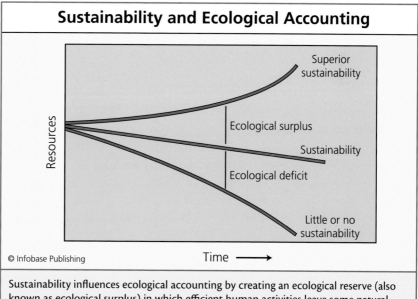

Sustainability influences ecological accounting by creating an ecological reserve (also known as ecological surplus) in which efficient human activities leave some natural resources untouched. Nonsustainable activities use up resources faster than the Earth can replenish them, and this creates an ecological deficit.

Renewable resources can be overexploited just as nonrenewable resources can if the use-up rate outpaces the replenishment rate. For example, the ivory trade of the 1800s wiped out entire elephant communities in Africa and Asia and decimated the numbers of elephants remaining in the wild. Not until 1989 did a ban on the worldwide ivory trade begin to save declining elephant herds. (Today poachers continue to target elephants' tusks.) Chemists have developed plastics to mimic the qualities of ivory and conserve this precious substance.

Environmental economics consist of three different philosophies on natural resources. The first group of economists, neoclassical economists, views natural resources as limitless because people will invent new technologies to replace these resources before they can be depleted. These economists believe the world's economy may also grow in a limitless manner. The second group consists of ecological economists who state that limitless economic growth is not sustainable in a world containing limited resources. Unlimited economic growth therefore leads to resource overexploitation. A third philosophy belongs to environmental economists who agree that some forms of natural-resource use cannot continue unless new technologies replace nonrenewable resources. The new technologies would allow people to consume resources almost as fast as they do now.

The following quotations provide insight into these three economic philosophies as related to the environment.

- Neoclassical economist Milton Friedman: "The question is, what is the supply curve of energy? The use of coal or oil is simply a means of producing energy. The stock of coal, of oil, etc., is certainly in some sense finite, but that doesn't mean that the potential amount of energy capable of being produced by whatever source is to be considered finite. Energy will be produced in any way that is cheapest at the time and as new means of producing energy are discovered the particular mode of producing energy will change from coal to oil to natural gas to atomic sources."

- Ecological economist Robert Costanza: "Historically, the recognition by humans of their impact upon the Earth has consistently lagged behind the magnitude of the damage they have imposed, thus weakening efforts to control this damage. Even today, technological optimists and others ignore the mounting evidence of global environmental degradation until it intrudes more inescapably upon their personal welfare."

- Environmental economists Michelle Haefele and John Loomis: "As the economic issues surrounding the environment and natural resources have grown beyond 'all or nothing' polarized debates such as 'growth or no growth,' economics has risen more and more to the forefront of the scientific information used by conservation groups. Some of this attention has surrounded the concept of sustainability. Conservation groups generally recognize that the production of commodities will not cease altogether and that, in order to achieve multiple goals, this production needs to be done in a way that is sustainable."

Time may show that each of these views on the environment and economics offers a reasonable argument and each probably also holds flaws that are as yet undiscovered.

CULTURAL INDICATORS OF SUSTAINABILITY

Viewpoints on the environment range between two extremes. Some people believe the Earth's resources exist for humans to use as they see fit. People at the other extreme feel humans are tiny cogs in the universe who do not own natural resources. Both opinions, and all of the shades in between, should accept that sustainability will help humans and not hurt them. Communities successful at building sustainability therefore strive for certain features, shown in the following list. These represent the major cultural indicators of sustainability:

- working with businesses rather than against them to solve environmental problems
- shifting emphasis from cleanup to prevention and minimization of things harmful to the environment

- dismantling strident win-at-all-costs groups and emphasizing the common goals of environmentalists and industrialists
- using scientific facts for making decisions rather than emotions or exaggerated information

Sustainability depends a great deal on people in different positions in society who are willing to work together on environmental issues that affect everyone. The thought that "we're all in this together" creates powerful teamwork in solving some very pressing problems.

ENVIRONMENTAL INDICATORS

The state of the environment can be evaluated by indicators much the same way countries measure economic indicators. A country's gross domestic product (GDP) comprises the total annual market value of all the goods and services produced by all the businesses in the country. The GDP provides a general indication of a country's relative wealth or lack of wealth. However, impoverished subpopulations hide in even the wealthiest countries, so another indicator has been developed to make a better estimate of environmental progress. This value is called the *genuine progress indicator* (GPI), which evaluates the factors used in calculating GDP but includes additional components such as the value of volunteer work and the costs of environmental damage. Redefining Progress is an economic *think tank* in Washington, D.C., that created the GPI to enable "policymakers at the national, state, regional, or local level to measure how well their citizens are doing both economically and socially." The public seldom hears of the GPI, but international organizations have increasingly used this tool to gauge threats to the environment.

Specific environmental indicators relate directly to air quality, water quality, toxic chemicals and their health risks, the status of the land and forests, and climate change brought about by greenhouse gas emissions. For example, the World Bank and other international organizations use the environmental indicators listed in Appendix E.

The World Economic Forum in 2008 compiled a ranking of countries based on economic indicators similar to the World Bank's. Of 149 countries to receive a ranking, the United States ranked 39th; its low ranking resulted mainly from poor air quality. The ranking report's lead author,

Daniel Esty of the Yale Center for Environmental Law and Policy, pointed out that "the country's performance on a new indicator that measures regional smog is at the bottom of the world right now." The *New York Times* reporter Felicity Barringer wrote, "Christine Kim, a research associate of Professor Esty's, calculated that a country's wealth, measured as GDP per capita, tended to correlate with a strong performance on such indicators as sanitation, indoor air quality and success in combating diseases—but also with a poor performance on greenhouse gas emissions and agricultural policies." According to Esty and his colleagues, it seems likely that economics and environment will be connected to each other into the future.

Some environmentalists have become increasingly critical of the relationships people such as Esty have drawn between national economics and sustainability. Ron Nielson, the author of *The Little Green Handbook* said, "GDP is an illusion of progress, a disguise that hides the stark reality of an environmental overdraft. We borrow from the future to pay for the present, and we use GDP to make ourselves happy." The GPI has not yet been universally accepted by economists. If economists and environmentalists come to agreement on how GPI can be used, it may well become an important indicator of sustainability.

CHALLENGES TO SUSTAINABILITY

The significant challenge for people who try to build sustainable communities comes from the incredible amount of information that science still lacks about the Earth. Each year environmental scientists publish thousands of technical papers on new insights into how ecosystems work, yet scientists may have identified only 10 percent or less of all species. That means that unknown organisms are carrying out undefined roles each second of each minute in soil, ponds, the ocean, remote jungles, and so forth. The failed experiment of Biosphere 2 shows the hazards of trying to create a natural environment solely from human input. Scientists would do better by mimicking what they know about nature and working with nature to build sustainability. The following major routes to achieving this sustainability may also represent humanity's greatest challenge:

- establish biodiversity protections that work
- build meaningful ways to decrease poverty

- prevent and reduce wastes and pollution
- decrease dependence on fossil fuels and increase dependence on renewable energy
- stop overconsumption; simplify lifestyles
- control population size and growth rate
- redirect economic systems to sustain resources rather than exploit resources

CASE STUDY: LESSONS FROM BIOSPHERE 2

In the desert ringing Tucson, Arizona, sits a 3.14-acre (1.27 ha), glass-paneled structure of 7.2 million cubic feet (204,000 m³) that is a popular tourist site. In 1991 and 1994 this structure called Biosphere 2 carried out one of the country's most innovative ecology experiments with human subjects. On September 26, 1991, eight scientists entered the Biosphere facility for a two-year stint in practicing sustainability. The biospherians, as they were called, lived in the enclosed structure that contained its own air supply and water-generation systems, and they grew their own food. Biosphere 2 held five different ecosystems: a simulated ocean with a coral reef; mangrove wetlands; a tropical rain forest; savannah grassland; and a fog desert. The critical question to be answered by Biosphere 2 centered on whether the ecosystems would generate the oxygen, water, and climate needed to sustain the biospherians.

All of the components built into Biosphere 2 had been designed to recreate the Earth's natural nutrient cycling. The ocean water would evaporate, condense to rain, and nourish the vegetation. Excess rain would flow through the wetland that would purify it into drinking water. Wastes received biological treatment. In addition to the eight people, Biosphere 2 contained more than 4,000 organisms that were expected to create an ecosystem that could support the scientists for two years or longer.

Biosphere 2 soon demonstrated an undeniable fact: Nature builds sustainable systems better than humans build them. Soil microbes began almost at once to consume oxygen and emit CO_2 inside the building. The CO_2 accumulation disrupted natural carbon cycling, and Biosphere 2's nitrogen cycle then malfunctioned. The health of sensitive animal species declined, and about 20 species went extinct. With the extinctions, other animals dominated ecosystems, which further harmed the health of the surviving animals and plants. For example, all the plants that depended on pollination for propagation died because Biosphere 2's pollinating insects could not survive.

All of these challenges remind scientists and students that sustainability is not easy to come by in society. Though every other species on Earth seems to have learned how to create sustainable lifestyles, humans have a hard time doing it. Sustainable communities can only be built if people rethink their needs and adjust their behavior toward new products, fast cars, and other signs of affluence. Until that happens, the more knowledge scientists accumulate on how ecosystems work, the better everyone's chances at building a sustainable community.

The lessons science gathered from Biosphere 2 turned out to be quite the opposite of what had been planned. Instead of discovering ways to build sustainable systems from scratch, the project's scientists discovered how inept humans can be in trying to mimic the Earth's environment. The *Discover* magazine writer Stephen Ornes revisited Biosphere 2's problems in a 2007 article: "After most crops failed, the team lived on emergency rations from outside the bubble until a depleting oxygen supply brought the experiment to a halt. 'Basically, we suffocated, starved, and went mad,' said biospherian Jane Poynter in 2003." The Biosphere 2 scientists

The Biosphere 2 experiment showed that people would have an extremely difficult time replacing ecosystems on their own if nature could no longer play this role. Despite the vast knowledge that ecologists have accumulated in the past century, far more questions remain about how nature builds and repairs sustainable ecosystems. *(Anthony Blake)*

Joel Cohen and David Tilman explained, "No one yet knows how to engineer systems that provide humans with life-supporting services that natural ecosystems provide for free." Cohen and Tilman's opinion hints at the obstacles in front of communities that wish to rebuild sustainable systems modeled on nature.

PERMACULTURE

Permaculture is short for permanent agriculture, which is the design and maintenance of ecosystems that grow an agricultural product while retaining a healthy ecosystem. The three main characteristics of healthy ecosystems are biodiversity, stability, and resilience. Permaculture also requires that the new ecosystem work with nature's landscape and maintain as much natural vegetation as possible. Permaculture differs from commercial agriculture in that commercial farming exists for the purpose of selling all of its output for people's use. Permaculture owners, by comparison, grow only the amount of crop that a community needs.

Permacultures differ based on the type of land they cover. Mountainsides require different techniques than rangeland, for example, but certain factors appear in almost all permacultures:

- plantings in polyculture (having different plant species in one crop)
- plantings of perennial species
- natural pollination
- natural plants, animals, insects, and fungi
- designed for food production, garden, pasture, or woodland
- farm animals allowed to forage on fallen fruit, vegetables, and plants
- manure tilled into the soil where it lies
- worm composting
- landscape provides rainwater catchment and storage
- surrounded by undisturbed natural habitat

Permaculture farms work in areas with poor soils or in places where the climate does not favor good plant growth if owners incorporate three key features: (1) hedge plants to attract pollinators and provide wildlife shelter; (2) plants grown specifically to attract wildlife or farm-animal foraging to help till the soil; and (3) plants that absorb and sequester soil nutrients and return them to the soil upon decay.

Innovative permaculture layouts have accommodated space for animals such as poultry, pollinating insects, and orchard trees, even within big cities. Each town and city in the United States differs in rules on keeping animals and setting up other types of agriculture. The Urban Permaculture Guild, based in California, has explained its idea on blending permaculture with city life: "Once we look at the city's elements as opportunities

Urban Permaculture Farm

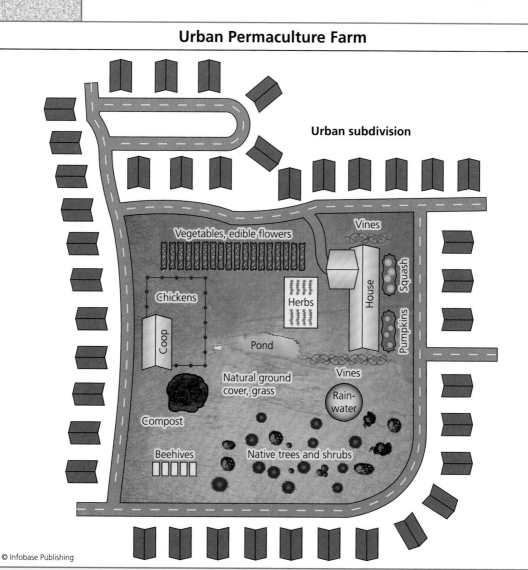

Urban subdivision

Vegetables, edible flowers

Vines

Squash

House

Pumpkins

Chickens

Herbs

Coop

Pond

Vines

Natural ground cover, grass

Rain-water

Compost

Beehives

Native trees and shrubs

© Infobase Publishing

An urban permaculture farm can take many shapes depending on climate, the owner's objectives, and local regulations on agriculture within city limits. The example shown here raises animals as well as crops. Permaculture farms typically include rainwater collection to conserve surface or underground sources, ground cover plants that reduce runoff and erosion, native plants and trees to maintain the local animal and plant life, and on-site waste treatment, such as compost areas.

to create beauty through design, we begin to find solutions, including: building greenhouses and planting gardens in vacant lots. Turning our grass lawns into gardens that produce food and are aesthetically beautiful. Creating community by combining backyards with our neighbors; tearing down fences to create bigger parcels of usable land. Using the presence of buildings to cultivate plants that need partial shade or vertical climbing space. Designing a water catchment system on the roofs of our houses that can provide all the water needs for the garden." People wishing to bring some of these features onto their property need to check with local statutes.

Several cities in Australia and Britain have begun undertaking permaculture. Mal McKenna wrote in 2008 for the Permaculture Research Institute of Australia on clever ways to fit sustainable agriculture into a city setting: "In a small area, the main concern is space. Following the recommendations of your local agricultural inspector or the directions on seed packets might leave just enough room for one small tree and a lettuce plant in a small backyard. However, small areas can be intensively planted as they can be (relatively) intensely cared for. Getting it all in is a matter of going up, going down, going sideways, and going with the flow." Permaculture therefore requires an open and creative mind.

McKenna's advice for urban permaculture consists of the following four choices: (1) "going up" means building hanging trellises or high fences to support vines and other climbing plants, and also beehives, gardens, or water catchments on roofs; (2) "going down" refers to getting the most out of soil moisture by the close-planting of vegetation that has deep roots next to plants that have shallow roots, as well as the addition of a small pond or a terraced garden to maximize the use of water; (3) "going sideways" encompasses an intense polyculture in as much horizontal space as possible, including ways to reuse plants as food, mulch, or a soil improver; and (4) "going with the flow" is the philosophy of using what the land offers rather than trying to mold the land to people's desires.

Owners of urban permaculture properties have found inventive ways to work with land and plants. For example, a poor-growing tree can act as a trellis, acidic soil can be devoted to growing berries, and a small pond can fill a low spot that is too moist for gardening.

In the United States, permaculture has gathered a slow but steady following in regions of Oregon and the Southwest, and a growing list of individual cities (Santa Fe, New Mexico, and Austin, Texas) have explored urban permaculture. Permaculture cannot serve large numbers of people, but it certainly contributes to the overall sustainability of communities.

CONCLUSION

Sustainable communities that successfully conserve natural resources depend on the combined cooperation of local government, residents, and inventors of new technologies. Sustainable communities may not be a mere luxury; these communities may represent the new way that cities and towns must live to keep society from using up all the world's natural resources.

Eco-cities that have been attempted, and perhaps others being planned, should take all their residents into consideration—the wealthy, the middle classes, and the poor. By doing this, a community can raise its living standards and its attention to the environment at the same time. In fact, community economics will become a bigger part of all future planning in environmental issues. Since the birth of the current environmental movement, most people have come to realize that environmental programs work best when they work with businesses rather than against the business community. The best hope for the environment in the future therefore ties in with industry profit. Many environmental scientists have used traditional economic indicators such as GDP to assess environmental well-being.

Sustainable communities will increasingly show that protections to the environment must include the input of all of the following groups, not just one or two: federal government, local government, community residents, city planners, sociologists, ecologists, and engineers. A successful sustainable community will result from technology, leadership, and the public working together.

FUTURE NEEDS

Sustainability seems to be an easy matter at first glance, but as communities delve into plans for sustainability, they often uncover difficult obstacles. Sustainability is a far more complex endeavor than simply recycling bottles and reducing waste. Conservation of resources requires a combination of many factors and a balance between technology, social values, and leadership.

Technology alone cannot create new sustainable uses of natural resources unless society makes adjustments in the way it views the world's natural capital. Some of these adjustments entail fairly easy recycling programs. Other adjustments consist of more difficult tasks such as changing the way people commute, giving up wasteful products, and seeking alternatives for nonrenewable natural resources. Most difficult, sustainability requires the solution of numerous problems that have been harming the environment for decades, and these problems have been extremely difficult to tackle in the past. Industrialized nations may prove to be a more challenging place to introduce sustainability than undeveloped parts of the world. This is because people in industrialized society have grown used to their conveniences, many of which demand large quantities of energy and nonrenewable resources.

To be successful, sustainability programs must confront things such as poverty, the way industry has operated for 100 years, and the uncontrolled growth of a human population in parts of the world that can no longer support the people they already have. Individual nations are hard-pressed to take these steps on their own, especially in places where centuries-old customs, religion, or warring factions control each citizen. Countries such as the United States must accept a leadership role, but many nations today do not trust U.S. leaders. For this reason, respected, apolitical international organizations offer the best chance of steering the world toward sustainable actions.

Is sustainability realistic? Some sectors have been successful in applying the basics of sustainability: the recycling industry; new chemicals; biodegradable materials; and biotechnology that seeks to invent new substances as well as conserve endangered species. Breakthroughs in green, white, and blue biotechnology for the environment have been limited so far, but their current innovations and those that are underway suggest that, yes, sustainability is realistic.

If technologies for building sustainability exist, why do sustainable ways of doing things enter society so slowly? Much of the reason rests in the need to understand the environment in order to affect the environment, and science still holds a massive amount of unknowns regarding life on land and in the ocean. Large international research programs to small laboratory experiments all serve a purpose in building a store of knowledge on the environment. This takes time. Leaders and the public must be reminded on occasion that scientific inquiry develops on its own schedule. New information about the universe comes only after a large volume of indisputable evidence has amassed.

While scientists work in laboratories and collect data in field studies, ordinary people have plenty of opportunities to build sustainability. Grassroots organizations have led the way in stirring interest and alarm over environmental issues. People can influence environmental programs with their votes, through volunteer work, and by giving money to important environmental causes. These actions actually describe the very manner in which today's respected environmental movement began.

Environmental causes have strong allies today because politicians—with a few regrettable exceptions—have made the environment an important part of their platforms. This certainly differs from the treatment given environmentalists in the 1960s and even later, when being an environmentalist opened a person to ridicule. Perhaps science is not the only thing that moves slowly; society can also be slow to accept change. Unfortunately, the status of many of today's natural resources warns that people do not have the luxury of time to fix the environment.

Each new U.S. administration has an opportunity to establish an environmental philosophy and set up new environmental doctrines. A new administration can rebuild relationships with other countries and work with them and with international groups toward sustainability. Americans increasingly vote for leaders based on environmental issues, so the United States has an opportunity as never before to make the environment a priority. By voting for leaders who have feasible plans for sustainability, Americans can make a tremendous difference in global environmental decisions.

Sustainable use of natural resources defines the environment's biggest need. Sustainability in turn requires a monumental effort to control the population growth rate so that people no longer overshoot their ecological footprints as they are now doing. World leaders, world health organizations, and educators must create a workable plan for controlling population, however difficult this may be.

Technology's future needs are highlighted by the following: development of bioengineered items that conserve resources in a safe manner; alternatives for wood, oil, and other hard-to-replace items; chemistry to create new biodegradable materials; further understanding of ecosystems; expanded uses for nanotechnology; and the discovery of materials in nature that can substitute for nonrenewable materials.

People like the idea of doing things that help the environment, but as Kermit the Frog once correctly warned, "It's not easy being green." Sustainability comes from a hundred big decisions and millions of small choices. To create a sustainable Earth, towns, countries, and continents will all take part. Sustainability may turn out to be history's first truly cooperative effort from everyone alive today.

Appendix A

MILESTONES IN BIOTECHNOLOGY'S HISTORY	
DATE	**EVENT**
4000–2000 B.C.E.	Egyptians use yeast to leaven bread and ferment beer; Babylonians engineer date palms with selective breeding
500 B.C.E.	China uses antibiotic from soybean mold to treat boils
100 C.E.	China uses insecticide made from chrysanthemums
1590	Janssen invents microscope
1663	Hooke describes cells
1675	Leeuwenhoek discovers bacteria
1797	Jenner develops a smallpox vaccine
1830	proteins discovered
1833	first enzyme discovered
1857	Louis Pasteur proposes microorganisms cause fermentation
1865	Gregor Mendel discovers genetic traits are passed from one generation to its offspring
1879	William James Beal develops hybrid corn

(continues)

MILESTONES IN BIOTECHNOLOGY'S HISTORY
(continued)

Date	Event
1915	bacterial viruses called phages are discovered
1919	the word *biotechnology* used for the first time in print
1928	Alexander Fleming discovers penicillin
1933	hybrid corn is commercialized
1944	Oswald Avery and Colin MacLeod show DNA carries genetic information
1946	genetic material from different viruses combined to make a new virus
1947	Barbara McClintock discovers transposons or "jumping genes" that move from place to place on a DNA molecule
1949	Linus Pauling shows sickle cell anemia is caused by a mutation
1953	James Watson and Francis Crick publish a description of the DNA molecule
1956	Kornberg discovers an enzyme needed for DNA replication
1963	Norman Borlaug develops high-yield wheat
1966	the genetic code is identified
1967	automated protein sequencers arrive
1971	laboratory synthesis of a gene
1973	Stanley Cohen and Herbert Boyer perfect techniques to creating new DNA out of pieces of DNA
1976	DNA sequencing begins; first biotechnology company (Genentech) founded

1977	a human gene is expressed in bacteria
1978	recombinant insulin produced
1980	gene synthesizing begins
1981	first transgenic animal produced in the United States; first animal clone produced in China
1983	an entire plant, the petunia, grown from biotechnology; polymerase chain reaction invented
1985	DNA fingerprinting results used in a courtroom; pest-resistant bioengineered plants grown in the field
1987	bioengineered food, virus-resistant tomatoes, tested in the field
1989	bacteria used to clean up oil spills
1990	gene therapy cures an immune disorder
1992	FDA declares bioengineered foods to be safe
1995	full gene sequence for bacteria identified
1997	18 bioengineered crops approved for use in the United States
1998	full gene sequence for an animal (worm) identified
2002	full gene sequence of humans published
2003	bioengineered crops used in 18 countries; first successfully cloned animal, Dolly the sheep
2004	United Nations declares bioengineered crops as valid way to alleviate hunger
2005	genetically engineered crops reach 222 million acres (89.8 million ha) worldwide
2008	plant fibers genetically modified for easier biofuel production

Appendix B

GLOBAL INDUSTRIES WITH A FUTURE IN WHITE BIOTECHNOLOGY	
INDUSTRY	**APPLICATIONS**
chemicals	bulk chemicals; synthesis; bioplastics; biopolymers
construction	road-surface components; biopolymer structures; oil-well construction materials
energy	biological fuel cells; bioethanol fuel; oil and gas desulfurization; biologically produced hydrogen
food	flavorings; oils; vitamins; enzymes; sweeteners; bread making; vegetable-oil processing
metals	biopolymer replacement for car components; electroplating; metal cleaning; ore mining
pharmaceuticals	drugs; contact lens solutions; nutritional supplements
pulp and paper	biological pulping; paper bleaching
textiles	synthetic fibers; stonewashing; textile dewatering
waste treatment	wastewater treatment; methane recovery; biodegradation
Source: Biotechnology Industry Organization	

Appendix C

U. S. MARINE BIOTECHNOLOGY INSTITUTES AND ORGANIZATIONS		
ORGANIZATION	**BIOTECHNOLOGY EMPHASIS**	**WEB SITE, URL (ACCESSED MARCH 27, 2009)**
Center of Marine Biotechnology, University of Maryland	conservation and enhancement of marine and estuarine species	http://www.umbi.umd.edu/comb/home.php
Center of Marine Biotechnology and Biomedicine, University of California, San Diego	new drugs and materials for human medicine, bioremediation, and specialized fields	http://cmbb.ucsd.edu
National Institute for Undersea Science and Technology, University of Mississippi	development of new technologies for medicines or agriculture and for marine research	http://www.niust.org
National Oceanic and Atmospheric Administration Undersea Research Program	marine-derived materials and ocean biodiversity conservation	http://www.nurp.noaa.gov/Biotech.htm
Scripps Institution of Oceanography	ocean ecosystem research	http://sio.ucsd.edu
Woods Hole Oceanographic Institution	sensors for new materials, genes, or as pollution detectors	http://www.whoi.edu

Appendix D

THE HISTORY OF PLASTICS AND POLYMERS		
DATE	INVENTOR	EVENT
1835	Henri Victor Regnault, French organic chemist	first to experiment with long molecules called hydrocarbons; first to make polyvinyl chloride
1862	Alexander Parkes, British chemist	discovered a moldable organic material extracted from cellulose
1869	John Wesley Hyatt, American inventor	used an extract from laurel trees to develop the first thermoplastic, a plastic that is molded and shaped under heat and pressure ** and the precursor to celluloid
1891	Louis Marie Hilaire Bernigaut, French textile developer	invented rayon by modifying an extract from cellulose to mimic silk
1900	Jacques Edwin Brandenberger, Swiss textile engineer	developed the first cellophane
1907	Leo Baekeland, American chemist	developed a liquid resin that composed the first thermoset plastics, materials that, once molded, would continue to hold the same shape, and made woods more durable and lightweight
1913	Jacques Edwin Brandenberger	added rayon to cellophane to produce a clear, protective, waterproof wrapping material

1920–1939	Wallace Hume Carothers, American chemist working at DuPont	developed nylon stockings
1926	Waldo Semon, American organic chemist working at B. F. Goodrich	developed commercially useful vinyl and polyvinyl chloride
1933	Ralph Wiley, American chemist working at Dow Chemical	discovered polyvinylidene chloride (Saran wrap)
1933	E. W. Fawcett and R. O. Gibson, American organic chemists	developed polyethylene
1938	Roy Plunkett, American chemist at DuPont	discovered Teflon
1940s	H. Staudinger, German chemist	identified the structure of plastic molecule
1940s	various chemical companies	developed neoprene, acrylic, polyethylene, and other synthetic polymers for plastic-making
1945–1949	various chemical companies	polyethylene becomes first plastic selling more than 1 billion pounds (4.5 million kg) per year; remains the highest-selling plastic
1949	James Wright, American engineer at General Electric	mixed silicone oil with boric acid to invent Silly Putty
1957	George de Maestral, Swiss engineer	used nylon strands to invent Velcro
1976	chemical industry	plastic became the most used material in the world

** Hyatt's invention also served to save thousands of elephants that were being killed for their tusks used in making billiard balls.

Source: American Chemistry Council

Appendix E

MAJOR ENVIRONMENTAL INDICATORS	
MAIN CATEGORY	**INDICATORS**
population	total world population urban population, percent of total gross domestic product (GDP) gross national income (GNI) per capita
agriculture	land area agricultural land, percent of total land area irrigated land, percent of cropland fertilizer consumption rural population density
emissions and pollution	CO_2 per unit of GDP CO_2 per capita particulate matter passenger cars per 1,000 people
energy	GDP per energy use energy use per capita energy from combustible waste energy imports, percent of use energy power consumption per capita energy generated by coal, percent of total
environmental health	respiratory infections in children diarrhea in children under-five mortality rate per 1,000 live births

forests and biodiversity	forest area, percent of total land area
	annual deforestation
	national protected areas, percent of total land area
	total known mammal species
	total threatened mammal species
	total known bird species
	total threatened bird species
national accounting and economics	gross savings
	consumption of fixed resources
	education expenditure
	energy depletion
	mineral depletion
	forest depletion
water and sanitation	freshwater resources per capita
	freshwater withdrawal
	access to improved water sources
	access to sanitation

Source: The World Bank

Glossary

AGROECOLOGY   science of using ecological concepts in the design and management of sustainable farms.

AGROECOSYSTEM   a farm or other agriculture area that acts as an ecosystem.

AQUACULTURE   raising of particular fish species in a confined area rather than in the open ocean.

AQUAPONICS   blending of agricultural methods for raising new plants and aquaculture methods for rearing the plants in water.

ARTIFICIAL INSEMINATION   the application of male semen to female ova by medical methods rather than by natural breeding.

BAGASSE   plant residue after a product, such as juice, has been extracted.

BIOCAPACITY BUFFER   land set aside to maintain healthy ecosystems and species and not used for human activities.

BIOCATALYSIS   driving a reaction forward using biological factors, such as enzymes, rather than physical or chemical factors, such as heat or acids, respectively.

BIOCOMPOSTABLE   biologically derived products that have been formulated to break down quickly in the environment after use.

BIODEGRADABLE   property of a substance to break down in nature due to biological and physical activities.

BIODIVERSITY   variety of different species, genes within a species, or different ecosystems.

BIOENGINEERING   process of developing an organism with new traits by putting genes from another species into the organism's DNA.

BIOETHANOL   alcohol derived from plants and used as a fuel alternative to fossil fuel.

BIOFUEL   any energy-generating fuel made from biological materials rather than fossil fuel.

BIOME   any region on Earth characterized by certain types of animal life, vegetation, and climate.

BIOMONITORING   the analysis of the body's blood, tissue, and other components for the presence of environmental chemicals.

BIOPLASTIC   material derived from plants or animals and used as a fuel alternative to chemical plastics.

BIOPOLYMER   long, chainlike molecule derived from a biological source.

BIOSENSOR   a device containing a biological component and a chemical component for detecting compounds in nature.

BIOSPHERE   the part of planet Earth containing life.

BIOTA   all Earth's living things.

BLUE BIOTECHNOLOGY   biotechnology based on marine materials or for the purpose of improving marine conditions.

BROAD-SPECTRUM CHEMICAL HERBICIDES   type of chemical pesticide that kills a wide variety of weeds.

CARRYING CAPACITY   maximum population size of a species that a habitat or an ecosystem can support over time without degrading the environment.

CHROMOSOME   the entire genetic material of an organism, made up mainly of DNA.

COMPANION FARMING   annual crops that are planted on land containing perennial crops, such as grass, during the perennial plant's dormant months.

CONSUMPTION LAND USE   all the land, fishing grounds, and materials that are used to provide people with food, shelter, mobility, and products.

CRYOPRESERVATION   the long-term storage of biological materials, such as microbes or tissue, at extremely cold temperature.

CYANOBACTERIA   bacteria that conduct photosynthesis for producing energy.

DEEP ECOLOGY   manner of caring for the Earth by viewing humans as part of nature with needs that are neither more significant nor less significant than all other organisms.

DIATOMS   unique hard-shelled algae that belong to the broad group of plankton species.

DINOFLAGELLATES   unique algae that swim by use of one or more tails (flagella) and that belong to the broad group of plankton species.

ECO-CITY   urban area that has maximized all ways it can conserve its resources.

ECOLOGICAL DEFICIT   condition when humanity's demand on nature is less than the biosphere's capacity to supply natural resources.

ECOLOGICAL FOOTPRINT   calculation of how much water and land a population needs to produce the resources it consumes and to degrade the wastes it produces.

ECOLOGICAL OVERSHOOT   condition when humanity's demand on nature exceeds the biosphere's capacity to supply or replace natural resources.

ECOLOGICAL RESERVE   natural resources available when a population's ecological footprint does not exceed the land's capacity to replenish the resources.

ECOSYSTEM   a community of species interacting with one another and with the nonliving things in a certain area.

ECOSYSTEM MODELING   use of miniature scale models of real ecosystems, such as the Chesapeake Bay, to study the effects of flooding, tides, runoff, erosion, and similar environmental activities.

ECOSYSTEM STABILITY   condition in which an ecosystem can recover from damage to it.

ENDOCRINE DISRUPTORS   chemicals that interfere with normal hormone function in the body.

ENTROPY   a state of disorder that contains insufficient energy needed to do work.

EUTROPHICATION   physical, chemical, and biological changes taking place in a body of water that has received sudden high levels of nutrients, usually nitrates and phosphates.

EXTREMOPHILE   microorganism that can grow in very harsh environments where most other organisms cannot live.

FEEDSTOCK   a raw material used to make another material or final product.

FOOD CHAIN   series of organisms in which each animal preys on the preceding animal in the chain and the size of each species usually increases toward the top of the chain.

FOOD WEB   complex network of interconnected food chains and feeding relationships.

FRAGILE ECOSYSTEM   an interrelated community of few species or a community that occupies a habitat that is easily destroyed.

FREE ENERGY OF ACTIVATION   an amount of energy that a chemical reaction must overcome before the reaction can go forward.

GENE BARCODING (DNA barcoding)   use of a specific unique gene or set of genes to identify a species, related organisms, or an individual organism.

GENETICALLY MODIFIED ORGANISM (GMO)   microbe, plant, or animal that contains one or more genes from another species.

GENOMICS   the study of all the genes in a single animal or species.

GENUINE PROGRESS INDICATOR (GPI)   a calculation for assessing the economic strength of a nation, using GDP plus the estimated value of volunteer work and the estimated cost of environmental damage.

GLOBAL HECTARE   unit of land area that puts a value on biocapacity; a hectare equals 2.47 acres.

GRASSROOTS   small, local organizations that create change in society.

GREENHOUSE GAS   any gas emitted into the atmosphere that holds heat in the lower atmosphere called the troposphere.

HABITAT   the place where a plant or animal lives.

INTEGRATED PEST MANAGEMENT   combination of methods for increasing the yield of healthy agricultural crops while inhibiting weeds and pests, with emphasis on natural or nonchemical methods whenever possible.

KINETICS   the science of motion or changing physical or chemical conditions.

MARKER GENE   a gene easy to identify and trace when transferring DNA from one organism to another organism.

MICROBES (microorganism)   microscropic living thing such as bacteria or protozoa.

NANOBIOTECHNOLOGY   the science of studying and using biological substances that are on a size scale measured in nanometers.

NATURAL CAPITAL   renewable and nonrenewable natural resources that sustain human populations on Earth.

NATURAL SELECTION   process by which one or more beneficial genes are reproduced in a species generation after generation.

NITROGEN FIXATION   a process whereby microbes absorb nitrogen gas from the atmosphere and convert the nitrogen to a form that can be used by plants.

NO-TILL FARMING   plant cultivation method in which growers leave plant residues in the field following a harvest and do not plow or disk the residues back into the soil.

NUTRIENT CYCLE (biogeochemical cycle)   the natural process of recycling the Earth's nutrients in various chemical forms through soil, water, air, and living things.

OFF-GAS   a vapor given off by a hazardous chemical, creating a health hazard if inhaled.

OLD-GROWTH FOREST   undisturbed trees that have resulted from a natural succession of plant and tree species and often can be hundreds of years old.

PERMACULTURE   design and maintenance of ecosystems that grow an agricultural product.

PHEROMONE   a substance that acts as a communication means between animals, usually by smell.

PHYTOREMEDIATION   water pollution cleanup using the action of plants and their root systems.

PLANKTON   tiny organisms that supply nutrients to larger and more complex marine organisms.

PLANT-INCORPORATED PROTECTANT (PIP)   a pesticide or other compound made by a plant to help the plant survive, often created through bioengineering.

POLYMER   a long, chainlike chemical compound.

POLYSACCHARIDE   a long molecule composed completely or mainly of sugars.

PRIMARY RECYCLING (closed-loop recycling)   process of recycling a material or product to make more of the same material or product.

RAPIDLY RENEWABLE WOOD   fast-growing trees that supply wood that serves as an alternative to wood from rare or endangered trees.

RECOMBINANT DNA   DNA formed with genes from two different and unrelated organisms.

SALTWATER INTRUSION   movement of salt water into freshwater sources.

SECONDARY RECYCLING (downcycling)   process of recycling a material or product to make a new product.

SPECTROSCOPY   process of identifying compounds by measuring the light spectrum they produce.

SUICIDE GENE   a gene inserted into a genetically engineered organism for the purpose of initiating destruction of the organism after a period of time in the environment.

SUSTAINABILITY   the ability of any system to survive for a given period of time.

THINK TANK   a group of experts that evaluates government policies and publishes opinions and advice on those policies.

TOXIN   a poison made by a microbe, plant, or animal.

TRANSFORMATION   process of inserting DNA from one organism into another organism.

TREE HUGGER   a term, often derogative, for an environmentalist.

VOLATILE ORGANIC COMPOUNDS (VOCs)   carbon-containing compounds that evaporate into the air.

WATER CYCLE   the natural process of recycling the Earth's water in various forms through soil, water, the atmosphere, and living things.

Further Resources

PRINT AND INTERNET

Australia Department of Agriculture and Food. "Farming for the Future: Demonstrating Sustainability." Available online. URL: www.agric.wa.gov.au/content/SUST/F4Fbrochure1.pdf. Accessed December 28, 2008. A brochure that describes Australia's sustainable farming.

Baharuddin, Hj. Ghazali. "Timber Certification: An Overview." Available online. URL: http://nzdl.sadl.uleth.ca/cgi-bin/library?e=d-00000-00---off-0fi1998--00-0--0-10-0---0---0prompt-10---4-------0-11--11-en-50---20-about---00-0-1-00-0-0-11-1-0utfZz-8-00&a=d&c=fi1998&cl=CL1.22&d=HASHa5daad7d45532b94892e3b. Accessed December 28, 2008. A detailed explanation of timber certification in Malaysia.

Balz, Dan, and Juliet Eilperin. "Gore and U.N. Panel Share Peace Prize." *Washington Post* (10/13/07). Available online. URL: www.washingtonpost.com/wp-dyn/content/article/2007/10/12/AR2007101200364.html. Accessed December 28, 2008. News article covering the announcement that Al Gore shared the 2007 Nobel Peace Prize.

Barringer, Felicity. "U.S. Given Poor Marks on the Environment." *New York Times* (1/23/08). Available online. URL: www.nytimes.com/2008/01/23/washington/23enviro.html?_r=1&scp=2&sq=Environment&st=nyt&oref=slogin. Accessed December 28, 2008. A news article provides an update on countries' environmental performance.

Bartram, William. *Travels through North and South Carolina, Georgia and Florida.* Philadelphia, Pa.: James and Johnson, 1791. Interesting first-person accounts of an early environmentalist.

BBC News. "Biofuel Use 'Increasing poverty.'" (6/25/08). Available online. URL: http://news.bbc.co.uk/2/hi/europe/7472532.stm. Accessed December 27, 2008. A short review of the disadvantages caused by biofuels.

Berg, Paul. "Dissections and Reconstructions of Genes and Chromosomes." Lecture presented at the Nobel Prize award ceremony, Stockholm, Sweden, December 8, 1980. Available online. URL: http://nobelprize.org/nobel_

prizes/chemistry/laureates/1980/berg-lecture.pdf. Accessed December 28, 2008. Berg recounts the history of genetic engineering.

Bradley, Robert L., Milton Friedman, and Richard L. Stroup. "Friedman's Legacy for Freedom and the Environment." *PERC Reports* 25, no. 3 (2007): 21–23. Available online. URL: www.perc.org/articles/article903.php. Accessed December 28, 2008. A retrospective on economist Friedman's views on how to manage environment and natural resources.

Brahic, Catherine. "Giant Microwave Turns Plastic Back to Oil." *New Scientist* (6/26/07). Available online. URL: http://environment.newscientist.com/article/dn12141. Accessed December 28, 2008. A short article that explains the technology behind recycling plastics into its original constituents.

Business Wire. "The Future of White Biotechnology is Attractive in China, and Considerable Progress in White Biotechnology Has Been Made in Recent Years." News release (5/19/08). Available online. URL: http://findarticles.com/p/articles/mi_m0EIN/is_2008_May_19/ai_n25433772. Accessed December 28, 2008. A short update on the opportunities evolving for white biotechnology in China.

Cape Cod Times. "Before We Harvest . . ." (08/11/08). Available online. URL: www.capecodonline.com/apps/pbcs.dll/article?AID=/20080811/OPINION/808110318. Accessed December 28, 2008. An editorial on the flaws in the George W. Bush administration's zoning regulations for ocean aquaculture.

Carson, Rachel. *Silent Spring.* Boston, Mass.: Houghton Mifflin, 1962. The groundbreaking book that awakened the public to chemicals in the environment.

Central Intelligence Agency. *The World Factbook 2008.* Washington, D.C.: Central Intelligence Agency, 2008. Available online. URL: www.cia.gov/library/publications/download/index.html. Accessed December 28, 2008. An annual compilation of global conditions on health, environment, and additional subjects.

Childress, James J., Horst Felbeck, and George N. Somero. "Symbiosis in the Deep Sea." *Scientific American* (May 1987). An in-depth article on the strange lifeforms that inhabit deep-sea thermal vents.

Cole, Thomas. "Lament of the Forest." *Knickerbocker* 17, no. 6 (1841), in *The Environmental Debate.* Edited by Peninah Neimark and Peter Rhoades Mott. Westport, Conn.: Greenwood Press, 1999.

Costanza, Robert, John Cumberland, Herman Daly, Robert Goodland, and Richard Norgaard. "Introduction to Ecological Economics." Chap. 1 in *The Encyclopedia of Earth,* edited by Nancy E. Golubiewski and Cutler J. Cleveland. Washington, D.C.: Environmental Information Coalition, National Council for Science and the Environment, 2007. A resource on ecology covering a broad range of subjects.

Cowley, Geoffrey. "Biotechnology: DNA on the Dinner Table." *Newsweek* (01/19/01). Available online. URL: www.mindfully.org/GE/Dinner-Table. htm. Accessed December 29, 2008. This article recounts the mishaps that have taken place in bioengineering prior to 2001.

CropLife International. "Agricultural Biotechnology Critical for Biodiversity Protection." Press release (03/14/06). Available online. URL: www.croplife.org/ library/attachments/49c06f6a-989e-49d6-abf1-28dea92e8e7c/2/2006%200 3%2014%20-%20Agricultural%20biotechnology%20critical%20for%20 biodiversity%20protection.pdf. Accessed December 28, 2008. An industry press release highlighting the benefits of bioengineered crops.

DeBare, Ilana. "Ridding World of Plastic Forks." *San Francisco Chronicle* (1/7/07). Available online. URL: www.sfgate.com/cgi-bin/article.cgi?file=/chronicle/ archive/2007/01/07/BUG8KNE27Q1.DTL&type=business. Accessed December 29, 2008. An article covering an entrepreneur's idea for biodegradable eating utensils and plates.

———. "Green Product Seals Are Gray Area." *San Francisco Chronicle* (4/19/08). Available online. URL: www.sfgate.com/cgi-bin/article.cgi?f=/c/ a/2008/04/19/MNHGVQQIC.DTL. Accessed December 29, 2008. An article discussing industry's attempts to use environmental claims as a marketing tool.

———. "Fishery Crisis Leads to New Approaches." *San Francisco Chronicle* (8/10/08). Available online. URL: http://socialissues.wiseto.com/Articles/ 182569857. Accessed December 29, 2008. An article covering the regulation and issues surrounding aquaculture.

Erickson, Brent. *New Biotech Tools for a Cleaner Environment.* Washington, D.C.: Biotechnology Industry Organization, 2004. Available online. URL: www. bio.org/ind/pubs/cleaner2004. Accessed December 30, 2008. Clear and detailed overview of green and white biotechnologies.

Frazzetto, Giovanni. "White Biotechnology." *EMBO Reports.* Available online. URL: www.nature.com/embor/journal/v4/n9/full/embor928.html. Accessed December 28, 2008. Excellent overview of the goals and the applications of white biotechnology.

Gibson, Elizabeth. "Costs Make Recycling Bottles a Tough Sell." *Columbus Dispatch* (8/4/08). Available online. URL: www.columbusdispatch.com/live/ content/local_news/stories/2008/08/04/recycle_glass.ART_ART_08-04- 08_B1_88AU95P.html?sid=101. Accessed December 28, 2008. An article on the poor economics of recycling glass compared with other recyclable materials.

Gladwell, Malcolm. *The Tipping Point.* New York: Back Bay Books, 2002. A popular book on the manners in which small actions cause major effects.

Gore, Al. *An Inconvenient Truth.* New York: Rodale, 2006. The former vice president's landmark book on the current state of the environment, especially global warming.

———. "Al Gore's Challenge to Repower America." Speech given at the Conference of Parties-14, Poznan, Poland, December 12, 2008. Available online. URL: www.repoweramerica.org. Accessed December 29, 2008. Al Gore encourages specific environmental objectives in a speech to a climate change conference.

Gorman, Tom. "Biotech Critic Considers Darker Questions of Science." *Los Angeles Times* (5/29/91). Available online. URL: www.foet.org/press/articles/BC/Los%20Angeles%20Times%20May%2029,%201991.pdf. Accessed December 29, 2008. An article that discusses the early safety concerns concerning bioengineered organisms.

Greenpeace. "Fed Farmers Foist Pro-GE Lobbyist on NZ Grain Growers." Press release (3/24/04). Available online. URL: www.greenpeace.org/new-zealand/press/releases/fed-farmers-foist-pro-ge-lobby. Accessed December 29, 2008. The environmental group's viewpoint on the disadvantages of bioengineered crops.

Grinnell, George, and Charles Sheldon, eds. *Hunting and Conservation.* New Haven, Conn.: Yale University Press, 1925. Historical background on natural resource conservation.

Grossman, Jeffrey. "Nanotechnology Takes Off." *KQED-TV Quest* (3/27/07). Available online. URL: www.kqed.org/quest/television/view/189?gclid=CJz-huOdi5UCFSQbagodsEYbrA. Accessed December 29, 2008. A video describing the potential uses of nanotechnology as discussed by experts from the University of California.

Haefele, Michelle, and John Loomis. "The Expanding Role of Environmental Economists in Conservation Organizations." Association of Environmental and Resource Economists Newsletter 27, no. 1 (2007): 20–22. A discussion on the options for protecting environment by linking it to economic factors.

Hankonen, Johanna. "Out of the Forest." Translated by *Books from Finland* 4 (1998). Available online. URL: www.finlit.fi/booksfromfinland/bff/498/hankonen.htm. Accessed August 26, 2008. A detailed article on urban planning in Finland.

Harding, Stephan. "What Is Deep Ecology?" Schumacher College. Available online. URL: www.schumachercollege.org.uk/learning-resources/what-is-deep-ecology. Accessed December 29, 2008. A thoughtful discussion on deep ecology, a unique view of sustainability.

Hart, Maureen. *Guide to Sustainable Community Indicators,* 2nd ed. West Hartford, Conn.: Sustainable Measures, 1999. A guide for establishing community sustainability programs.

Hayward, Steven. *Index of Leading Environmental Indicators 2008.* San Francisco: Pacific Research Institute, 2008. Available online. URL: http://liberty. pacificresearch.org/docLib/20080401_08_Enviro_Index.pdf. Accessed December 29, 2008. Excellent global report on the current status of natural resources as well as toxic chemicals and health.

Hill, Gladwin. "Nation Set to Observe Earth Day." *New York Times* (4/21/70). Available online. URL: http://graphics8.nytimes.com/packages/pdf/topics/ earthday.pdf. Accessed December 29, 2008. A historical article introducing the first Earth Day.

Hoffman, Hillel. "Way Too Many for Us." *Cornell University Alumni News* (September 2004). Available online. URL: www.ecofuture.org/pk/pkcapcty.html. Accessed December 29, 2008. An argument for the need for population controls to conserve declining resources.

Johnston, David, and Kim Master. *Green Remodeling: Changing the World One Room at a Time.* Gabriola Island, British Columbia, Canada: New Society Publishers, 2004. Resource for alternative building materials.

Kahn, Jennifer. "Nano's Big Future." *National Geographic* (June 2006). An introduction to nanotechnology and possible applications.

Kunzig, Robert. "Pick Up a Mop." *Time* (7/14/08). A review of innovative ideas for removing carbon from the atmosphere.

Low, Nicholas, Brendan Gleeson, Ray Green, and Darko Radović. *The Green City: Sustainable Homes, Sustainable Suburbs.* London: Routeledge, 2005. A book with clear details on sustainable buildings and towns, including helpful resources.

Mabrey, Vicki, and Ely Brown. "Luxury Hotel Goes 'Green.'" ABC News (5/24/07). Available online. URL: www.theorchardgardenhotel.com/ images/press/ABC-News-Orchard-Garden.pdf. Accessed December 29, 2008. Transcript from a television news story that describes a San Francisco hotel working toward sustainability.

Manninen, Antti. "Espoo's Idealistic Model City Tapiola Turns Fifty." *Helsingin Sanomat International Edition* (5/8/03). Available online. URL: www2.hs.fi/ english/archive/news.asp?id=20030805IE3. Accessed December 29, 2008. Description of sustainable "garden city" in Finland.

Marks, Kathy, and Daniel Howden. "The World's Rubbish Dump: A Garbage Tip That Stretches from Hawaii to Japan." *Independent* (2/5/08). Available online. URL: www.independent.co.uk/environment/the-worlds-rubbish-dump-a-garbage-tip-that-stretches-from-hawaii-to-japan-778016.html. Accessed December 29, 2008. Description of an infamous collection of floating garbage that covers an expanse of the Pacific Ocean.

Martin, Marshall A., Jean R. Riepe, April C. Mason, and Peter E. Dunn. "Agricultural Biotechnology: Before You Judge." Purdue University Cooperative Extension Service document ID-201, 7/96. Available online. URL: www.ces.purdue.edu/extmedia/ID/ID-201-W.htm. Accessed December 29, 2008. A scholarly overview of the pros and cons of biotechnology in agriculture.

Masciangioli, Tina, and Wei-Xian Zhang. "Environmental Technologies at the Nanoscale." *Environmental Science and Technology* (3/1/03). Available online. URL: www.nano.gov/html/res/GC_ENV_PaperZhang_03-0304.pdf. Accessed December 29, 2008. Detailed and well-illustrated introduction to nanotechnology.

McKenna, Mal. "Transforming Your Urban Backyard." Permaculture Research Institute of Australia, 2008. Available online. URL: http://permaculture.org.au/2008/06/26/transforming-your-urban-backyard. Accessed December 27, 2008. Tips on building an urban permaculture landscape.

McKinsey and Company. *White Biotechnology: Gateway to a More Sustainable Future.* Brussels, Belgium: European Association for Bioindustries, 2003. Available online. URL: www.mckinsey.com/clientservice/chemicals/pdf/biovision_booklet_final.pdf. Accessed December 29, 2008. A booklet that explains the economic and energy-efficiency benefits of white biotechnology.

Miller, G. Tyler. *Environmental Science,* 11th ed. Belmont, Calif.: Thomson Learning, 2006. One of the best and comprehensive resources on environmental science.

Ministry of Science and Information and Communication Technology. *National Biotechnology Policy.* Dhaka, Bangladesh: Government of the People's Republic of Bangladesh, 2004. Available online. URL: www.promotebiotechbd.net/bb.pdf. Accessed December 29, 2008. A useful and concise overview of environmental biotechnology with a good glossary.

Monsanto Company. "Conversations about Plant Biotechnology." *Biotechnology Videos* (2006). Available online. URL: www.monsanto.com/biotech-gmo/asp/topic.asp?id=ConservationTillage#. Accessed December 29, 2008. A chemical company describes the benefits of plant biotechnology and no-tillage farming.

———. "Genetically Engineered Plants Deliver Significant Environmental and Economic Benefits." *Biotechnology Videos* (2008.) Available online. URL: www.monsanto.com/biotech-gmo/asp/experts.asp?id=RogerBeachy#mid. Accessed December 29, 2008. A plant pathologist discusses the advantages of bioengineered crops.

Morell, Virginia. "Way Down Deep." *National Geographic* (June 2004). Available online. URL: http://ngm.nationalgeographic.com/ngm/0406/feature2.

Accessed December 30, 2008. An article on the methods used by Monterey Bay Aquarium to study the deep ocean.

Mufson, Steven. "U.S. Oil Reserves Get a Big Boost." *Washington Post* (9/6/06). Available online. URL: www.washingtonpost.com/wp-dyn/content/article/2006/09/05/AR2006090500275.html. Accessed December 30, 2008. A news article on recent discoveries of new oil fields.

National Aquaculture Association. "Environmental Stewardship; Introduction: Challenge and Opportunity." Available online. URL: www.thenaa.net/environmental-stewardship. Accessed December 30, 2008. A professional aquaculture organization describes the benefits of this industry in the United States and mechanisms to ensure environmental safety.

Nielson, Ron. *The Little Green Handbook.* New York: Picador, 2006. A book that provides some unique viewpoints on the environment and useful data tables.

Obama, Barack. "Barack Obama's Acceptance Speech." *New York Times* (8/28/08). Available online. URL: www.nytimes.com/2008/08/28/us/politics/28text-obama.html?pagewanted=1&_r=1. Accessed December 28, 2008. The 2008 Democratic Convention speech that marked a historic national event in U.S. history.

Ogilvie, Felicity. "Penguin Droppings Help Identify Pesticide Hot Spots." ABC News (3/11/08). Available online. URL: www.abc.net.au/news/stories/2008/03/11/2186867.htm. Accessed December 30, 2008. Important findings on the spread of pesticides as far as the South Pole.

O'Neil, Caitlin. "How Architectural Salvage Yards Work." *This Old House Television.* Available online. URL: www.thisoldhouse.com/toh/article/0,,212818,00.html. Accessed December 30, 2008. A short report on the specialized business of salvaging valuable woods and fixtures from historical houses.

Organization for Economic Cooperation and Development. *Need for Research and Development Programs in Sustainable Chemistry.* Paris, France: OECD, 2002. Available online. URL: www.oecd.org/dataoecd/9/55/2079870.pdf. Accessed December 30, 2008. OECD's report on the status, as of the early 2000s, of sustainable chemistry relative to national governments, universities, and industry.

Ornes, Stephen. "Biosphere 2 Repurposed for Luxury Homes." *Discover* (9/21/07). Available online. URL: http://discovermagazine.com/2007/sep/buying-biosphere-2. Accessed December 30, 2008. A very short update on the current use of the Biosphere 2 facility.

Phillips, Tom. "Quiet Revolution." *Guardian* (3/26/08). Available online. URL: www.guardian.co.uk/society/2008/mar/26/communities.regeneration. Accessed

December 30, 2008. An article covering the possibilities for building urban sustainability, using Curitiba, Brazil, as an example.

Pollack, Andrew. "Kraft Recalls Taco Shells with Bioengineered Corn." *New York Times* (9/23/00). Available online. http://query.nytimes.com/gst/fullpage.html?res=9F0DEFD71F3BF930A1575AC0A9669C8B63&sec=&spon=&pagewanted=1. Accessed December 30, 2008. The news event in which a bioengineered plant product contaminated regular product and the ramifications.

PR Newswire Association. "Birthplace of the Green Movement Welcomes Newcomers." News release (6/17/08). Available online. URL: www.prnewswire.com/cgi-bin/stories.pl?ACCT=104&STORY=/www/story/06-17-2008/0004833958&EDATE=. Accessed December 27, 2008. A brief update on the current status of the U.S. green movement.

Public Broadcasting Service. "The Current Mass Extinction." WGBH Educational Foundation, 2001. Available online. URL: www.pbs.org/wgbh/evolution/library/03/2/l_032_04.html. Accessed December 30, 2008. An examination of current extinction rates and projections.

Redefining Progress. "Genuine Progress Indicator." Available online. URL: www.rprogress.org/sustainability_indicators/genuine_progress_indicator.htm. Accessed December 30, 2008. A simple explanation of the Genuine Progress Indicator as a better measure of sustainability than the Gross Domestic Product indicator.

Revkin, Andrew C. "Climate Experts Tussle Over Details. Public Gets Whiplash." *New York Times* (7/29/08). Available online. URL: www.nytimes.com/2008/07/29/science/earth/29clim.html?_r=1&scp=2&sq=global+warming&st=nyt&oref=slogin. Accessed December 30, 2008. This news article describes why scientists often debate the subject of global warming.

Royte, Elizabeth. "Corn Plastic to the Rescue." *Smithsonian* (August 2006). Available online. URL: www.smithsonianmag.com/science-nature/plastic.html. Accessed December 30, 2008. An in-depth article on biological sources for making biodegradable, compostable products.

Ruder, Kate. "Exploring the Sargasso Sea: Scientists Discover One Million New Genes in Ocean Microbes." Genome News Network (3/4/04). Available online. URL: www.genomenewsnetwork.org/articles/2004/03/04/sargasso.php. Accessed December 30, 2008. An article explaining methods used for genomics studies in the ocean.

Schneider, Keith. "Ideas and Trends; Man to Microbe: Do a Job, Drop Dead." *New York Times* (3/27/88). Available online. URL: http://query.nytimes.com/gst/fullpage.html?res=940DE1DE103EF934A15750C0A96E948260&sec=&spon=&pagewanted=1. Accessed December 30, 2008. This article describes the new, as of the 1980s, technology of suicide genes for bioengineered microbes.

Scripps Institution of Oceanography. "Scripps Expedition Provides New Baseline for Coral Reef Conservation." Scripps News (2/25/08). Available online. URL: http://scrippsnews.ucsd.edu/Releases/?releaseID=883. Accessed December 30, 2008. Methods used in monitoring coral reef health and the effect of climate change.

Shackley, Myra. *Wildlife Tourism*. London: Elsevier, 1996. A thoughtful review of potential effects of tourists on many types of animal life.

Shah, Sonia. *Crude: The Story of Oil*. New York: Seven Stories Press, 2004. An insightful book on the current global oil economy.

Sierra Club. "Biotechnology." *Sierra Club Conservation Policies* (2/20/01). Available online. URL: www.sierraclub.org/policy/conservation/biotech.aspx. Accessed December 30, 2008. The Sierra Club's official policy on genetic engineering.

Sijbesma, Feike. "From Charity to Development." Presented at DSM Special Session: BioVision. Lyon, France, March 11, 2007. Available online. URL: www.dsm.com/en_US/downloads/media/improving.pdf. Accessed December 30, 2008. A presentation that argues for sustainable practices in business.

Sisk, Taylor. "A Particular Concern: Protecting Deep-Sea Corals." NOAA's Undersea Research Program *In the Spotlight* (4/1/05). Available online. URL: www.nurp.noaa.gov/Spotlight/DeepSeaCorals.htm. Accessed December 30, 2008. A description of research methods and findings on coral reefs.

Sizer, Bridget Bentz. "Shrink Your Ecological Footprint." *Washington Post* (3/12/06). Available online. URL: www.washingtonpost.com/wp-dyn/content/article/2006/03/09/AR2006030902038.html. Accessed December 30, 2008. An article providing easy tips on how people can reduce their ecological footprints.

Skoloff, Brian. "Tire Reef Off Florida Proves a Disaster." *USA Today* (2/17/07). Available online. URL: www.usatoday.com/news/nation/2007-02-17-florida-reef_x.htm. Accessed December 30, 2008. An update on Florida's 1972 ill-fated attempt to build an artificial reef out of used tires.

Soller, Kurt. "Green Your Getaway." *Newsweek* (7/9/07). Available online. URL: www.theorchardgardenhotel.com/images/press/Newsweek.pdf. Accessed December 30, 2008. An article that presents environmentally friendly choices for vacations.

Steffen, Alex, ed. *Worldchanging: A User's Guide for the 21st Century*. New York: Harry N. Abrams, 2006. A useful handbook on numerous ways of building sustainability.

Stier, Ken. "Fish Farming's Growing Dangers." *Time* (9/19/07). Available online. URL: www.time.com/time/health/article/0,8599,1663604,00.html. Accessed December 29, 2008. An article on the environmental problems created by the aquaculture industry.

Stinson, Jeffrey. "Biofuel Push Questioned as Food Prices Soar." *USA Today* (4/24/08). Available online. URL: www.usatoday.com/money/industries/energy/2008-04-24-biofuels_N.htm. Accessed December 30, 2008. The economic drawbacks associated with biofuel from corn.

Strauss, Valerie. "Raining on Her Own Parade." *Washington Post* (4/21/08). Available online. URL: www.washingtonpost.com/wp-dyn/content/article/2008/04/20/AR2008042001423.html?sid=ST2008042002453. Accessed December 30, 2008. This article *reexamines* the importance of recent Earth Days.

Thompson, Larry. "Are Bioengineered Foods Safe?" *FDA Consumer* (January–February 2000). Available online. URL: www.fda.gov/fdac/features/2000/100_bio.html. Accessed December 30, 2008. A year 2000 review on the known safety effects of bioengineered food.

University of Texas. "New Source for Biofuels Discovered by Researchers at the University of Texas at Austin." Office of Public Affairs *News* (4/23/08). Available online. URL: www.utexas.edu/news/2008/04/23/biofuel_microbe. Accessed December 30, 2008. An article on microbial-based biofuels.

Venter, Craig. "Oceanic Expedition Uncovers Vast Genetic Diversity: Conversation with J. Craig Venter." By Ray Suarez. *The Online NewsHour,* PBS (3/16/07). Available online. URL: www.pbs.org/newshour/bb/science/jan-june07/ocean_03-16.html. Accessed December 30, 2008. A transcript of an interview with a renowned geneticist, discussing the enormous diversity of ocean organisms.

Walsh, Bryan. "The Truth about Plastic." *Time* (7/21/08). Available online. URL: www.time.com/time/magazine/article/0,9171,1821664,00.html. Accessed December 30, 2008. A report on the hazards of plastics and the compounds that compose plastics.

Williams, John, and Hester Gascoigne. "Redesign of Plant Production Systems for Australian Landscapes." Presented at the 11th Australian Agronomy Conference, Geelong, Victoria, Australia, February 2–6, 2003. Available online. URL: www.regional.org.au/au/asa/2003/i/4/williams.htm. Accessed December 30, 2008. A scholarly presentation on the progress of sustainable farming in Australia.

Williams, Ted. "As Ugly as a Tree." *Audubon* (September–October 2007). Williams provides an opinion on why indiscriminate tree-planting may harm the environment.

Wilson, Edward O., ed. *Biodiversity.* Washington, DC.: National Academy Press, 1988. A classic text on biodiversity, edited by the foremost expert on the subject.

Woods Hole Oceanographic Institution. "New Robot Sub Surveys the Deep off the Pacific Northwest." News release (8/8/08). Available online. URL: www.whoi.edu/page.do?pid=7545&tid=282&cid=47407&ct=162. Accessed December 30, 2008. An update and description of undersea exploration crafts.

World Wildlife Fund. "The Rising Tide of Fish Farming." (10/12/03). Available online. URL: www.panda.org/news_facts/newsroom/features/index.cfm?uNewsID=8281&uLangI D=1. Accessed December 30, 2008. The World Wildlife Fund's stance on aquaculture and the precautions suggested to protect the environment.

——. "Deep Sea: Importance." (2/29/08). Available online. URL: www.panda.org/about_wwf/what_we_do/marine/blue_planet/deep_sea/deepsea_importance/index.cfm. Accessed December 30, 2008. A short description of the value of marine materials to biotechnology.

Yollin, Patricia. "Group Brings City under the Canopy." *San Francisco Chronicle* (11/19/06). An article on the growing interest of replanting trees in highly populated cities.

Zabarenko, Deborah. "Global Warming May Cause World Crop Decline." Reuters (9/12/07). Available online. URL: www.cgdev.org/content/article/detail/14422. Accessed December 30, 2008. A special news report on global warming's effects on world food production.

WEB SITES

American Chemistry Council. Available online. URL: www.americanchemistry.com. Accessed December 10, 2008. Resource for environmental chemistry news and background on plastics.

Biotechnology Industry Organization. Available online. URL: www.bio.org. Accessed December 29, 2008. Reference for technical and regulatory topics concerning the biotechnology industry.

Consumer Reports Greener Choices. Available online. URL: www.greenerchoices.org. Accessed December 28, 2008. Resource for green consumer products, including cars, foods, electronics, and garden products, plus ecological news.

Earth Day Network. Available online. URL: www.earthday.net. Accessed December 28, 2008. Good resource on a variety of sustainability topics.

European Association for Bioindustries. Available online. URL: www.europabio.org. Accessed December 28, 2008. Superb resource on white biotechnology.

Food and Agriculture Organization. Available online. URL: www.fao.org. Accessed December 31, 2008. Comprehensive resource on environment and biodiversity related to world food supply; contains useful statistics on fishing and aquaculture.

Global Footprint Network. Available online. URL: www.footprintnetwork.org/ en/index.php/GFN. Accessed December 28, 2008. Basic primer on ecological footprints, with an excellent glossary.

Redefining Progress. Available online. URL: www.rprogress.org/index.htm. Accessed December 26, 2008. Explanations on economic-environmental relationships and public policy.

Sustainable Communities Network. Available online. URL: www.sustainable. org. Accessed December 28, 2008. A resource for community action on sustainability issues.

Urban Permaculture Guild. Available online. URL: www.urbanperma cultureguild.org. Accessed December 30, 2008. Useful information on establishing permaculture in individual cities.

World Bank. Available online. URL: www.worldbank.org. Accessed December 29, 2008. Resource for how economic plans address climate change, energy, and other broad areas of environmental science.

Index

Note: Page numbers in *italic* refer to illustrations. The letter *t* indicates tables.